# Android Native Development Kit Cookbook

A step-by-step tutorial with more than 60 concise recipes on Android NDK development skills

**Feipeng Liu**

BIRMINGHAM - MUMBAI

# Android Native Development Kit Cookbook

First published: March 2013

Production Reference: 1140313

Published by Packt Publishing Ltd.
Livery Place
35 Livery Street
Birmingham B3 2PB, UK.

ISBN 978-1-84969-150-5

www.packtpub.com

Cover Image by Artie Ng (artherng@yahoo.com.au)

# Credits

**Author**
Feipeng Liu

**Reviewers**
Roger Belk

Frank Grützmacher

Sylvain Ratabouil

**Acquisition Editor**
Martin Bell

**Commissioning Editor**
Shreerang Deshpande

**Lead Technical Editor**
Mayur Hule

**Technical Editors**
Lubna Shaikh

Worrell Lewis

**Project Coordinator**
Leena Purkait

**Proofreader**
Aaron Nash

**Indexer**
Monica Ajmera Mehta

**Graphics**
Aditi Gajjar

Valentina D'silva

**Production Coordinator**
Shantanu Zagade

**Cover Work**
Shantanu Zagade

# About the Author

**Feipeng Liu** is a technology enthusiast who is focused on multimedia systems and applications. He started mobile applications development in 2008 on Windows Mobile. Since Feb 2010, he has been developing apps for Android with NDK. His Android apps have been used by many users. One of his apps, **video converter Android**, has reached one million downloads within 10 months. Feipeng received his B.ENG in Electrical and Electronic Engineering degree from Nanyang Technological University, and Master of Computing degree in the Department of Computer Science from National University of Singapore.

I would like to thank Shreerang Deshpande for offering me the opportunity to author this book and for helping me throughout the writing, and Leena Purkait for keeping track of its progress and coordination, Mayur Hule, Lubna Shaikh, and Worrell Lewis for editing the book drafts! I would like to express my gratitude to other Packt Publishing staff who helped with the book! My grateful thanks are also extended to Roger, Frank and Sylvain, who offered great suggestions during the review.

I also would like to thank Assoc. Prof. Wei Tsang Ooi in National University of Singapore, the supervisor of my master project. A lot of stuff in this book is based on the things I learnt during the project.

Last but not least, I would like to thank my parents Zhulan Shen and Yi Liu, and Ms. Yang Xiaoqing for the support and understanding during the writing of the book.

# About the Reviewers

**Roger Belk** is a 45-year-old self-taught Android developer with 20 + apps in Google's Play Store under the developer name **BigTexApps**. He started out using Google's & MIT's App Inventor and then after two years of learning to use AI, he taught himself to use Java to build Android apps. He has reviewed two other books, *Google App Inventor, Ralph Roberts, Packt Publishing* (ISBN 978-1-84969-212-0) and *Android 3.0 Animation, Alex Shaw, Packt Publishing* (ISBN 978-1-84951-528-3).

**Frank Grützmacher** spent some years in the research of distributed electronic design tools and worked for several German blue chip companies such as Deutsche Post and AEG. He was involved in Android platform extensions for a mobile manufacturer. Therefore, on one hand he knows how to build large enterprise apps and on the other hand how to make Android system apps.

He is currently working for the IT daughter of the largest German Telco company.

In the past, he has reviewed Corba- and Java-related books for American and German publishers.

**Sylvain Ratabouil** is a confirmed IT consultant experienced with C++ and Java technologies. He worked in the space industry and got involved in aeronautic projects at Valtech, where he is now taking part in the digital revolution.

As a technology lover, he is passionate about mobile technologies and cannot live or sleep anymore without his Android smart phone.

# www.PacktPub.com

## Support files, eBooks, discount offers and more

You might want to visit www.PacktPub.com for support files and downloads related to your book.

Did you know that Packt offers eBook versions of every book published, with PDF and ePub files available? You can upgrade to the eBook version at www.PacktPub.com and as a print book customer, you are entitled to a discount on the eBook copy. Get in touch with us at service@packtpub.com for more details.

At www.PacktPub.com, you can also read a collection of free technical articles, sign up for a range of free newsletters and receive exclusive discounts and offers on Packt books and eBooks.

http://PacktLib.PacktPub.com

Do you need instant solutions to your IT questions? PacktLib is Packt's online digital book library. Here, you can access, read and search across Packt's entire library of books.

## Why Subscribe?

- Fully searchable across every book published by Packt
- Copy and paste, print and bookmark content
- On demand and accessible via web browser

## Free Access for Packt account holders

If you have an account with Packt at www.PacktPub.com, you can use this to access PacktLib today and view nine entirely free books. Simply use your login credentials for immediate access.

# Table of Contents

# Preface

Since its first release in 2008, Android has become the largest mobile platform in the world. The total number of apps in Google Play is expected to reach 1,000,000 in mid 2013. Most of the Android apps are written in Java with Android Software Development Kit (SDK). Many developers only write Android code in Java even though they are experienced with C/C++, without realizing what a powerful tool they are giving up.

Android **Native Development Kit** (**NDK**) was released in 2009 to help developers write and port native code. It offers a set of cross-compiling tools and a few libraries. Programming in NDK offers two main advantages. Firstly, you can optimize your apps in native code and boost performance. Secondly, you can reuse a large number of existing C/C++ code. *Android Native Development Kit* is a practical guide to help you write Android native code with NDK. We will start with the basics such as **Java Native Interface** (**JNI**), and build and debug a native app (chapter 1 to 3). We will then explore various libraries provided by NDK, including OpenGL ES, Native Application API, OpenSL ES, OpenMAX AL, and so on (Chapters 4 to 7). After that, we will discuss porting existing applications and libraries to Android with NDK (Chapters 8 and 9). Finally, we will demonstrate how to write multimedia apps and games with NDK (Bonus chapters 1 and 2).

## What this book covers

*Chapter 1, Hello NDK*, covers how to set up an Android NDK development environment in Windows, Linux, and MacOS. We will write a "Hello NDK" application at the end of the chapter.

*Chapter 2, Java Native Interface*, describes the usage of JNI in detail. We will call native methods from the Java code and vice versa.

*Chapter 3, Build and Debug NDK Applications*, demonstrates building native code from a command line and Eclipse IDE. We will also look at debugging native code with gdb, cgdb, eclipse, and so on.

*Chapter 4, Android NDK OpenGL ES API*, illustrates OpenGL ES 1.x and 2.0 APIs. We will cover 2D drawing, 3D graphics, texture mapping, EGL, and so on.

*Chapter 5, Android Native Application API*, discusses Android native application APIs, including managing native windows, accessing sensors, handling input events, managing assets, and so on. We will see how to write a pure native app in this chapter.

*Chapter 6, Android NDK Multithreading*, depicts Android multithreading API. We will cover creating and terminating native threads, various thread synchronization techniques (mutex, conditional variables, semaphore, and reader/writer lock), thread scheduling, and thread data management.

*Chapter 7, Other Android NDK API*, discusses a few more Android libraries, including `jnigraphics`, the dynamic linker library, the `zlib` compression library, the OpenSL ES library, and the OpenMAX AL library.

*Chapter 8, Porting and Using Existing Libraries with Android NDK*, describes various techniques of porting and using existing C/C++ libraries with NDK. We will port the `boost` library at the end of the chapter.

*Chapter 9, Porting Existing Applications to Android with NDK*, provides a step-by-step guide for porting an existing application to Android with NDK. We use an open source image resizing program as an example.

*Bonus Chapter 1, Developing Multimedia Applications with NDK*, demonstrates how to write multimedia applications with the `ffmpeg` library. We will port the `ffmpeg` library and use the library APIs to write a frame grabber application.

*Bonus Chapter 2, Developing Games with NDK*, discusses writing games with NDK. We will port the Wolfenstein 3D game to show how to set up game display, add game control, and enable audio effects for a game.

You can download the bonus chapters from `http://www.packtpub.com/sites/default/files/downloads/Developing_Multimedia_Applications_with_NDK.pdf` and `http://www.packtpub.com/sites/default/files/downloads/Developing_Games_with_NDK.pdf`.

# What you need for this book

A computer with Windows, Ubuntu Linux, or MacOS installed is necessary (Linux or MacOS is preferable). Although we can run Android apps with an emulator, it is slow and inefficient for Android development. Therefore, it is recommended to have an Android device.

The book assumes a basic understanding of C and C++ programming languages. You should also be familiar with Java and Android SDK.

Note that the sample code of this book is based on Android ndk r8 unless otherwise stated, since it is the latest version of NDK at the time of writing. By the time the book is published, there should be newer versions. The code should also run on any newer versions. Therefore we can install NDK r8 or later.

# Who this book is for

The book is written for anyone who is interested in writing native code for Android. The chapters are arranged from basic to intermediate to advanced, and they are relatively independent. Readers who are new to NDK are recommended to read from the beginning to the end, while readers who are familiar with NDK can pick up any specific chapters or even specific recipes.

# Conventions

In this book, you will find a number of styles of text that distinguish between different kinds of information. Here are some examples of these styles, and an explanation of their meaning.

Code words in text are shown as follows: "Windows NDK comes with a new `ndk-build.cmd` build script."

A block of code is set as follows:

```
#include <string.h>
#include <jni.h>

jstring
Java_cookbook_chapter1_HelloNDKActivity_naGetHelloNDKStr(JNIEnv* pEnv,
jobject pObj)
{
    return (*pEnv)->NewStringUTF(pEnv, "Hello NDK!");
}
```

When we wish to draw your attention to a particular part of a code block, the relevant lines or items are set in bold:

```
LOCAL_PATH := $(call my-dir)
include $(CLEAR_VARS)
LOCAL_MODULE   := framegrabber
LOCAL_SRC_FILES := framegrabber.c
#LOCAL_CFLAGS := -DANDROID_BUILD
LOCAL_LDLIBS := -llog -ljnigraphics -lz
LOCAL_STATIC_LIBRARIES := libavformat_static libavcodec_static
libswscale_static libavutil_static
include $(BUILD_SHARED_LIBRARY)
$(call import-module,ffmpeg-1.0.1/android/armv5te)
```

Any command-line input or output is written as follows:

```
$sudo update-java-alternatives -s <java name>
```

**New terms** and **important words** are shown in bold. Words that you see on the screen, in menus or dialog boxes for example, appear in the text like this: "Go to **Control Panel | System and Security | System | Advanced system settings**."

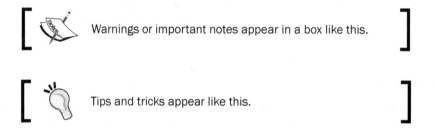

Warnings or important notes appear in a box like this.

Tips and tricks appear like this.

# Reader feedback

Feedback from our readers is always welcome. Let us know what you think about this book—what you liked or may have disliked. Reader feedback is important for us to develop titles that you really get the most out of.

To send us general feedback, simply send an e-mail to feedback@packtpub.com, and mention the book title via the subject of your message.

If there is a topic that you have expertise in and you are interested in either writing or contributing to a book, see our author guide on www.packtpub.com/authors.

# Customer support

Now that you are the proud owner of a Packt book, we have a number of things to help you to get the most from your purchase.

## Downloading the example code

You can download the example code files for all Packt books you have purchased from your account at http://www.packtpub.com. If you purchased this book elsewhere, you can visit http://www.packtpub.com/support and register to have the files e-mailed directly to you.

## Errata

Although we have taken every care to ensure the accuracy of our content, mistakes do happen. If you find a mistake in one of our books—maybe a mistake in the text or the code—we would be grateful if you would report this to us. By doing so, you can save other readers from frustration and help us improve subsequent versions of this book. If you find any errata, please report them by visiting http://www.packtpub.com/submit-errata, selecting your book, clicking on the **errata submission form** link, and entering the details of your errata. Once your errata are verified, your submission will be accepted and the errata will be uploaded on our website, or added to any list of existing errata, under the Errata section of that title. Any existing errata can be viewed by selecting your title from http://www.packtpub.com/support.

## Piracy

Piracy of copyright material on the Internet is an ongoing problem across all media. At Packt, we take the protection of our copyright and licenses very seriously. If you come across any illegal copies of our works, in any form, on the Internet, please provide us with the location address or website name immediately so that we can pursue a remedy.

Please contact us at copyright@packtpub.com with a link to the suspected pirated material.

We appreciate your help in protecting our authors, and our ability to bring you valuable content.

## Questions

You can contact us at questions@packtpub.com if you are having a problem with any aspect of the book, and we will do our best to address it.

# 1
# Hello NDK

In this chapter, we will cover the following recipes:

- ▶ Setting up an Android NDK development environment in Windows
- ▶ Setting up an Android NDK development environment in Ubuntu Linux
- ▶ Setting up an Android NDK development environment in Mac OS
- ▶ Updating Android NDK
- ▶ Writing a Hello NDK program

## Introduction

Android NDK is a toolset that allows developers to implement a part of or an entire Android application in a native language, such as C, C++, and assembly. Before we start our journey to NDK, it is important to understand the advantages of NDK.

First of all, NDK may improve application performance. This is usually true for many processor-bound applications. Many multimedia applications and video games use native code for processor-intensive tasks.

The performance improvements can come from three sources. Firstly, the native code is compiled to a binary code and run directly on OS, while Java code is translated into Java byte-code and interpreted by Dalvik **Virtual Machine** (**VM**). At Android 2.2 or higher, a **Just-In-Time** (**JIT**) compiler is added to Dalvik VM to analyze and optimize the Java byte-code while the program is running (for example, JIT can compile a part of the byte-code to binary code before its execution). But in many cases, native code still runs faster than Java code.

Java code is run by Dalvik VM on Android. Dalvik VM is specially designed for systems with constrained hardware resources (memory space, processor speed, and so on).

The second source for performance improvements at NDK is that native code allows developers to make use of some processor features that are not accessible at Android SDK, such as NEON, a **Single Instruction Multiple Data** (**SIMD**) technology, allowing multiple data elements to be processed in parallel. One particular coding task example is the color conversion for a video frame or a photo. Suppose we are to convert a photo of 1920x1280 pixels from the RGB color space to the YCbCr color space. The naive approach is to apply a conversion formula to every pixel (that is, over two million pixels). With NEON, we can process multiple pixels at one time to reduce the processing time.

The third aspect is that we can optimize the critical code at an assembly level, which is a common practice in desktop software development.

 The advantages of using native code do not come free. Calling JNI methods introduces extra work for the Dalvik VM and since the code is compiled, no runtime optimization can be applied. In fact, developing in NDK doesn't guarantee a performance improvement and can actually harm performance at times. Therefore, we only stated that it may improve the app's performance.

The second advantage of NDK is that it allows the porting of existing C and C++ code to Android. This does not only speed up the development significantly, but also allows us to share code between Android and non-Android projects.

Before we decide to use NDK for an Android app, it is good to know that NDK will not benefit most Android apps. It is not recommended to work in NDK simply because one prefers programming in C or C++ over Java. NDK cannot access lots of APIs available in the Android SDK directly, and developing in NDK will always introduce extra complexity into your application.

With the understanding of the pros and cons of NDK, we can start our journey to Android NDK. This chapter will cover how to set up Android NDK development in Windows, Ubuntu Linux, and Mac OS. For developers who have set up an Android NDK development environment before, a recipe with detailed steps of how to update an NDK development environment is provided. At the end of the chapter, we will write a Hello NDK program with the environment setup.

# Setting up an Android NDK development environment in Windows

In this recipe, we will explore how to set up an Android NDK development environment in Windows.

## Getting ready

Check the Windows edition and system type. An Android development environment can be set up on Windows XP 32-bit, Windows Vista 32- or 64-bit, and Windows 7 32- or 64-bit.

Android development requires Java JDK 6 or above to be installed. Follow these steps to install and configure Java JDK:

1. Go to the Oracle Java JDK web page at `http://www.oracle.com/technetwork/java/javase/downloads/index.html`, and choose JDK6 or above for your platform to download.

2. Double-click on the downloaded executable, and click through the installation wizard to finish the installation.

3. Go to **Control Panel | System and Security | System | Advanced system settings**. A **System Properties** window will pop up.

4. Click on the **Environment Variables** button in the **Advanced** tab; another **Environment Variables** window will pop up.

5. Under **System variables**, click on **New** to add a variable with the name as JAVA_HOME and value as the path of the JDK installation root directory. This is shown as follows:

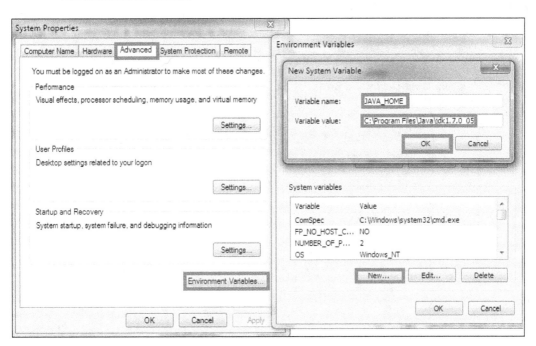

6. Under **System variables**, scroll to find the PATH (or Path) environment variable. Insert %JAVA_HOME%\bin; at the beginning of the value. If no PATH or Path variable exists, create a new variable with the value set to %JAVA_HOME%\bin. Click on **OK** all the way through to dismiss all windows.

7. To verify whether JDK is installed and configured correctly, start a new command-line console, and enter javac -version. If JDK is configured correctly, you will get the Java version in the output.

```
C:\Users\roman10>javac -version
javac 1.6.0_21
```

**Cygwin** is a Linux-like environment for Windows to run software available on Linux. Android NDK development requires Cygwin 1.7 or higher installed to execute some Linux programs; for example, the GNU make.

Since NDK r7, the Windows NDK comes with a new ndk-build.cmd build script, which uses NDK's prebuilt binaries for GNU make, awk, and other tools. Therefore Cygwin is not required for building NDK programs with ndk-build.cmd. However, it is recommended that you still install Cygwin, because ndk-build.cmd is an experimental feature and Cygwin is still needed by the debugging script ndk-gdb.

Follow these steps to install Cygwin:

1. Go to http://cygwin.com/install.html to download setup.exe for Cygwin. Double-click on it after the download is complete in order to start the installation.

2. Click on **Next**, then select **Install from Internet**. Keep clicking on **Next** until you see the **Available Download Sites** list. Select the site that is close to your location, then click on **Next**:

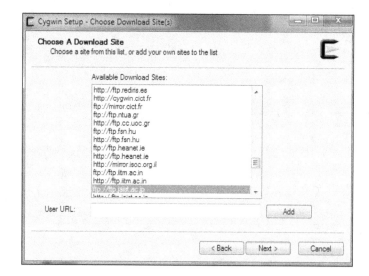

3.  Look for GNU **make** under **Devel**, ensure it is version 3.81 or later, and **gawk** under **Base**. Alternatively, you can search for make and gawk using the **Search** box. Make sure both GNU make and gawk are selected to install, then click on **Next**. The installation can take a while to finish:

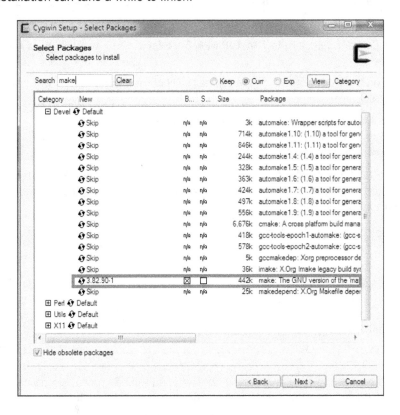

Eclipse is a powerful software **Integrated Development Environment** (**IDE**) with an extensible plugin system. It is the recommended IDE to develop Android apps. Go to http://www.eclipse.org/downloads/, and download the Eclipse Classic or Eclipse IDE for Java developers. Extract the compressed file and it will be ready for use. Note that Android development requires Eclipse 3.6.2 (Helios) or greater.

> The Android developer website provides an Android Developer Tools bundle at http://developer.android.com/sdk/index.html. It includes the Eclipse IDE with the ADT plugin, and the Android SDK. We can download this bundle and skip the SDK installation described in steps 1 to 10 of the following *How to do it...* section.

## How to do it...

The following steps show you how to set up an Android NDK development environment in Windows. We will first set up an SDK development environment. Steps 1 to 10 can be skipped if SDK is already set up.

1. Start Eclipse. Select **Help | Install New Software**, and a window titled **Install** will pop up.

2. Click on the **Add...** button at the top-right will corner, and another window titled **Add Repository** will pop up.

3. In the **Add Repository** window, enter ADT for **Name** and `https://dl-ssl.google.com/android/eclipse/` for **Location**. Then click on **OK**.

4. It may take a few seconds for Eclipse to load the software items from the ADT website. After loading, select **Developer Tools** and **NDK Plugins**, then click on **Next** to proceed:

| | |
|---|---|
| ☑ Developer Tools | |
| ☑ Android DDMS | 20.0.0.v201206242043-391819 |
| ☑ Android Development Tools | 20.0.0.v201206242043-391819 |
| ☑ Android Hierarchy Viewer | 20.0.0.v201206242043-391819 |
| ☑ Android Traceview | 20.0.0.v201206242043-391819 |
| ☑ Tracer for OpenGL ES | 20.0.0.v201206242043-391819 |
| ☑ NDK Plugins | |
| ☑ Android Native Development Tools | 20.0.0.v201206242043-391819 |

5. In the next window, a list of tools to be installed will be shown. Simply click on **Next**. Read and accept all the license agreements, then click on **Finish**.

6. After installation finishes, restart **Eclipse** as prompted.

7. Download Android SDK from `http://developer.android.com/sdk/index.html`.

8. Double-click on the installer to start the installation. Follow the wizard to finish the installation.

9. In Eclipse, select **Window | Preferences** to open the **Preferences** window. Select **Android** from the left panel, then click on **Browse** to locate the Android SDK root directory. Click on **Apply**, and then **OK**.

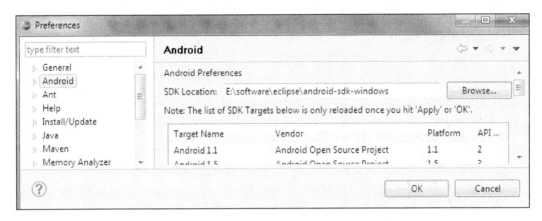

10. Start **Android SDK Manager** at the Android SDK installation root directory. Select **Android SDK Tools**, **Android SDK Platform-tools**, at least one Android platform (the latest is preferred), **System Image**, **SDK Samples**, and **Android Support**. Then click on **Install**. in the next window, read and accept all the license agreements, then click on **Install**:

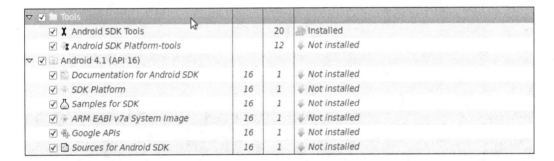

11. Go to `http://developer.android.com/tools/sdk/ndk/index.html` to download the latest version of Android NDK. Unzip the downloaded file.

**Downloading the example code**

You can download the example code files for all Packt books you have purchased from your account at `http://www.packtpub.com`. If you purchased this book elsewhere, you can visit `http://www.packtpub.com/support` and register to have the files e-mailed directly to you.

12. Open `Cygwin.bat` under the `cygwin` root directory. It contains the following content by default:

```
@echo off
C:
chdir C:\cygwin\bin
bash --login -i
```

13. Add the following content after `@echo off` before `C:`

```
set IS_UNIX=
set JAVA_HOME=<JDK path>
set PATH=<SDK path>\tools;<NDK path>
set ANDROID_NDK_ROOT=/cygdrive/<NDK path>
```

As an example, the file content on my machine is as follows; note that `Progra~1` is the short name for the `Program Files` folder:

```
set IS_UNIX=
set JAVA_HOME=c:/Progra~1/Java/jdk1.7.0_05
set PATH=C:/Users/Administrator/AppData/Local/Android/android-sdk/
tools;C:/Users/Administrator/Downloads/android-ndk-r8-windows/
android-ndk-r8
set ANDROID_NDK_ROOT=/cygdrive/c/Users/Administrator/Downloads/
android-ndk-r8-windows/android-ndk-r8
```

14. Start Cygwin by double-clicking on `cygwin.bat`, then go to the `samples/hello-jni` directory in NDK. Type the command `ndk-build`. If the build is successful, it proves that the NDK environment is set up correctly:

```
Administrator@2STS12S ~
$ cd /cygdrive/c/Users/Administrator/Downloads/android-ndk-r8-windows/android-n
dk-r8/samples/hello-jni/

Administrator@2STS12S /cygdrive/c/Users/Administrator/Downloads/android-ndk-r8-w
indows/android-ndk-r8/samples/hello-jni
$ ndk-build
Gdbserver      : [arm-linux-androideabi-4.4.3] libs/armeabi/gdbserver
Gdbsetup       : libs/armeabi/gdb.setup
Cygwin         : Generating dependency file converter script
Compile thumb  : hello-jni <= hello-jni.c
SharedLibrary  : libhello-jni.so
Install        : libhello-jni.so => libs/armeabi/libhello-jni.so
```

15. In Eclipse, select **Window | Preferences** to open the **Preferences** window. Click on **Android** from the left panel, and select **NDK** from the drop-down list. Click on **Browse** to locate the Android NDK root directory. Click on **OK** to dismiss the pop-up window. This enables us to build and debug Android NDK applications with the Eclipse NDK plugin:

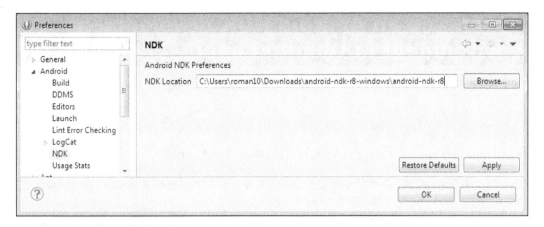

## How it works...

In this recipe, we first set up an Android SDK development environment and then the NDK development environment.

Android NDK does not require installation. We downloaded NDK, and configured the path to make it more convenient to use.

Cygwin is not required for Android SDK development, but is essential for NDK development because NDK uses some Linux tools that depend on Cygwin.

**NDK plugin in ADT:** NDK plugin for Eclipse is available in **Android Development Tools** (**ADT**), which allows us to build and debug Android NDK applications easily.

> The NDK plugin is only available for ADT 20.0.0 or later, which was released on June 2012. You may want to update your Eclipse ADT in order to use the NDK plugin.

## There's more...

We installed Eclipse IDE as a part of our development environment. Eclipse is the recommended IDE for developing Android applications, and it comes with lots of useful tools and utilities to help our development. However, it is not a compulsory component of the development environment.

# Setting up an Android NDK development environment in Ubuntu Linux

This recipe depicts how to set up an Android NDK development environment in Ubuntu Linux.

## Getting ready

Check your Ubuntu version and make sure it is version 8.04 or later.

GNU C Library (glibc) 2.7 or above is required. It is usually installed with Linux by default. Two simple methods can check the version of glibc:

1. Start a terminal, and enter `ldd --version`. This will print the version of `ldd` and `glibc`:

```
ldd (Ubuntu EGLIBC 2.13-20ubuntu5.1) 2.13
Copyright (C) 2011 Free Software Foundation, Inc.
This is free software; see the source for copying conditions.  There is NO
warranty; not even for MERCHANTABILITY or FITNESS FOR A PARTICULAR PURPOSE.
Written by Roland McGrath and Ulrich Drepper.
```

2. We can execute the library as an application. Start a terminal, locate the library location, and then enter the following command:

   `<glibc library location>/<glibc library>.`

   The following output will be displayed:

```
GNU C Library (Ubuntu EGLIBC 2.13-20ubuntu5.1) stable release version 2.13, by Roland McGrath et al.
Copyright (C) 2011 Free Software Foundation, Inc.
This is free software; see the source for copying conditions.
There is NO warranty; not even for MERCHANTABILITY or FITNESS FOR A
PARTICULAR PURPOSE.
Compiled by GNU CC version 4.6.1.
Compiled on a Linux 3.0.17 system on 2012-03-07.
Available extensions:
        crypt add-on version 2.1 by Michael Glad and others
        GNU Libidn by Simon Josefsson
        Native POSIX Threads Library by Ulrich Drepper et al
        BIND-8.2.3-T5B
libc ABIs: UNIQUE IFUNC
For bug reporting instructions, please see:
<http://www.debian.org/Bugs/>.
```

3. We need to enable 32-bit application execution if we are using a 64-bit machine. Start a terminal, and enter the following command:

   `sudo apt-get install ia32-libs`

4. Install JDK 6 or above. At a terminal, enter the command `sudo apt-get install openjdk-6-jdk`, or alternatively we can enter `sudo apt-get install sun-java6-jdk`. After installation, we need to add the JDK path to the `PATH` environment variable by adding the following lines to `~/.bashrc`:

```
export JDK_PATH=/usr/local/jdk1.7.0/bin
export PATH=$PATH:$JDK_PATH
```

We will use Eclipse as our IDE. Please refer to the *Setting up an Android NDK development environment in Windows* recipe for instructions.

## How to do it...

The following steps indicate the procedure of setting up an Android NDK development environment on Ubuntu Linux:

1. Follow steps 1 to 6 of the *Setting up an Android NDK development environment in Windows* recipe to install the ADT plugin for Eclipse.

2. Download Android SDK from `http://developer.android.com/sdk/index.html`, then extract the downloaded package.

3. Append the following lines to `~/.bashrc`:

```
export ANDROID_SDK=<path to Android SDK directory>
export PATH=$PATH:$ ANDROID_SDK/tools:$ANDROID_SDK/platform-tools
```

4. Follow steps 9 and 10 of the *Setting up an Android NDK development environment in Windows* recipe to configure the SDK path at Eclipse, and download additional packages.

5. Download the latest version of Android NDK from `http://developer.android.com/tools/sdk/ndk/index.html`, then extract the downloaded file.

6. Change the lines that you appended to `~/.bashrc` in step 3:

```
export ANDROID_SDK=<path to Android SDK directory>
export ANDROID_NDK=<path to Android NDK directory>
export PATH=$PATH:$ANDROID_SDK/tools:$ANDROID_SDK/platform-
tools:$ANDROID_NDK
```

7. Start a new terminal, then go to the `samples/hello-jni` directory in NDK. Type the command `ndk-build`. If the build is successful, it proves that the NDK environment is set up correctly:

```
                  :~/Downloads/android-ndk-r8/samples/hello-jni$ ndk-build
Gdbserver        : [arm-linux-androideabi-4.4.3] libs/armeabi/gdbserver
Gdbsetup         : libs/armeabi/gdb.setup
Compile thumb    : hello-jni <= hello-jni.c
SharedLibrary    : libhello-jni.so
Install          : libhello-jni.so => libs/armeabi/libhello-jni.so
```

## How it works...

We first set up Android SDK and then Android NDK. Ensure that the path is set properly, so that the tools can be accessed without referring to the SDK and NDK directories.

The `.bashrc` file is a startup file read by the bash shell when you start a new terminal. The export commands appended the Android SDK and NDK directory locations to the environment variable PATH. Therefore, every time a new bash shell starts, PATH is set properly for SDK and NDK tools.

## There's more...

The following are a few more tips on setting up an NDK development environment:

▶ **Configure Path at Startup File**: We append to the SDK and NDK paths to the PATH environment variable at `~/.bashrc` file. This assumes that our Linux system uses the bash shell. However, if your system uses another shell, the startup file used may be different. The startup files used by some commonly used shells are listed as follows:

  ❑ For C shell (`csh`), the startup file to use is `~/.cshrc`.

  ❑ For `ksh`, the startup file to use can be obtained using the command `echo $ENV`.

  ❑ For `sh`, the startup file to use is `~/.profile`. The user needs to log out of the current session and log in again for it to take effect.

▶ **Switch JDK**: In Android development, we can either use Oracle Java JDK or OpenJDK. In case we run into issues with any one of the JDKs, we can switch to another Java JDK, if we have installed both of them.

  ❑ To check which JDK the system is currently using, use the following command:

    `$update-java-alternatives -l`

  ❑ To switch between two JDKs, use the following command:

    `$sudo update-java-alternatives -s <java name>`

The following is an example for switching to Oracle JDK 1.6.0:

```
$sudo update-java-alternatives -s java-1.6.0-sun
```

# Setting up an Android NDK development environment in Mac OS

This recipe describes how to set up an Android NDK development environment in Mac OS.

## Getting ready

Android development requires Mac OS X 10.5.8 or higher, and it works on the x86 architecture only. Ensure that your machine meets these requirements before getting started.

Register an Apple developer account, then go to `https://developer.apple.com/xcode/` to download Xcode, which contains a lot of developer tools, including the `make` utility required for Android NDK development. After the download is complete, run the installation package and make sure that the **UNIX Development** option is selected for installation.

As usual, Java JDK 6 or above is required. Mac OS X usually ships with a full JDK. We can verify that your machine has the required version by using the following command:

```
$javac -version
```

## How to do it...

Setting up an Android NDK development environment on Mac OS X is similar to setting it up on Ubuntu Linux. The following steps explain how we can do this:

1. Follow steps 1 to 6 of the *Setting up an Android NDK development environment in Windows* recipe to install the ADT plugin for Eclipse.

2. Download Android SDK from `http://developer.android.com/sdk/index.html`, then extract the downloaded package.

3. Append the following lines to `~/.profile`. If the file doesn't exist, create a new one. Save the changes and log out of the current session:

   ```
   export ANDROID_SDK=<path to Android SDK directory>
   export PATH=$PATH:$ ANDROID_SDK/tools:$ANDROID_SDK/platform-tools
   ```

4. In Eclipse, select **Eclipse | Preferences** to open the **Preferences** window. Select **Android** from the left panel, then click on **Browse** to locate the Android SDK root directory. Click on **Apply**, and then **OK**.

5. In a terminal, start the Android SDK Manager at the `tools` directory by typing the command `android`. Select **Android SDK Tools, Android SDK Platform-tools**, at least one Android platform (the latest one is preferred), **System Image, SDK Samples**, and **Android Support**. Then click on **Install**. In the next window, read and accept all the license agreements, then click on **Install**.

6. Download the latest version of Android NDK from `http://developer.android.com/tools/sdk/ndk/index.html`, and then extract the downloaded file.

7. Change the lines that you appended to `~/.profile` in step 3:

```
export ANDROID_SDK=<path to Android SDK directory>
export ANDROID_NDK=<path to Android NDK directory>
export PATH=$PATH:$ANDROID_SDK/tools:$ANDROID_SDK/platform-
tools:$ANDROID_NDK
```

8. Start a new terminal, then go to the `samples/hello-jni` directory in NDK. Type the command `ndk-build`. If the build is successful, it proves that the NDK environment is set up correctly.

## How it works...

The steps to set up an Android NDK development environment on Mac OS X are similar to Ubuntu Linux, since both of them are Unix-like operating systems. We first installed Android SDK, then Android NDK.

# Updating Android NDK

When there is a new release of NDK, we may want to update NDK in order to take advantage of the new features or bug fixes with the new release. This recipe talks about how to update Android NDK in Windows, Ubuntu Linux, and Mac OS.

## Getting ready

Please read the previous recipes in this chapter, depending on the platform of your choice.

## How to do it...

In Windows, follow these instructions to update Android NDK:

1. Go to `http://developer.android.com/tools/sdk/ndk/index.html` to download the latest version of Android NDK. Unzip the downloaded file.

2. Open `Cygwin.bat` under the `cygwin` root directory. The content should be similar to the following code snippet, if you have previously configured NDK on the system:

```
@echo off
set IS_UNIX=
set JAVA_HOME=<JDK path>
set PATH=<SDK path>\tools;<NDK path>
set ANDROID_NDK_ROOT=/cygdrive/<NDK path>
C:
chdir C:\cygwin\bin
bash --login -i
```

3. Update `<NDK path>` from the old NDK path to the newly downloaded and decompressed location.

In Ubuntu Linux, follow these instructions to update Android NDK:

1. Download the latest version of Android NDK from `http://developer.android.com/tools/sdk/ndk/index.html`, then extract the downloaded file.

2. If we have followed the *Setting up an Android NDK development environment in Ubuntu Linux* recipe, the following content should appear at the end of `~/.bashrc`:

```
export ANDROID_SDK=<path to Android SDK directory>
export ANDROID_NDK=<path to Android NDK directory>
export PATH=$PATH:$ANDROID_SDK/tools:$ANDROID_SDK/platform-
tools:$ANDROID_NDK
```

3. Update the `ANDROID_NDK` path to the newly downloaded and extracted Android NDK folder.

In Mac OS, the steps are almost identical to Ubuntu Linux, except that we need to append the path to `~/.profile` instead of `~/.bashrc`.

## How it works...

NDK installation is completed by simply downloading and extracting the NDK file, and configuring the path properly. Therefore, updating NDK is as simple as updating the configured path to the new NDK folder.

## There's more...

Sometimes, updating NDK requires updating SDK first. Since this book focuses on Android NDK, explaining how to update SDK is beyond the scope of this book. You can refer to the Android developer website at `http://developer.android.com/sdk/index.html`, for details on how to do it.

At times, we may feel the need to use an old version of NDK to build certain applications because of compatibility issues. Therefore, it may be useful to keep multiple versions of Android NDK and switch between them by changing the path or simply using the full path to refer to a specific version of NDK.

# Writing a Hello NDK program

With the environment set up, let's start writing the code in NDK. This recipe walks through a Hello NDK program.

## Getting ready

The NDK development environment needs to be set up properly before starting to write the Hello NDK program. Please refer to previous recipes in this chapter, depending upon the platform of your choice.

## How to do it...

Follow these steps to write, compile, and run the Hello NDK program:

1. Start Eclipse, and select **File** | **New** | **Android Project**. Enter `HelloNDK` as the value for **Project Name**. Select **Create new project in workspace**. Then click on **Next**:

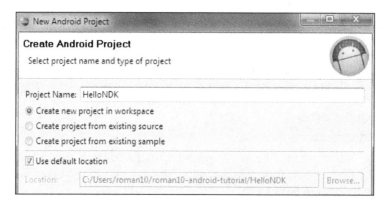

2. In the next window, select an Android version that you want to target. Usually, the latest version is recommended. Then click on **Next**.

3. In the next window, specify your package name as `cookbook.chapter1`. Select the **Create Activity** box, and specify the name as `HelloNDKActivity`. Set the value for **Minimum SDK** as `5 (Android 2.0)`. Click on **Finish**:

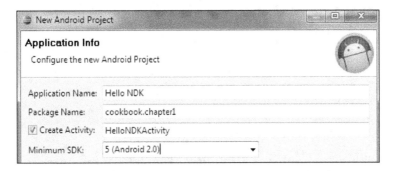

4. In the Eclipse package explorer, right-click on the `HelloNDK` project, and select **New | Folder**. Enter the name `jni` in the pop-up window, then click on **Finish**:

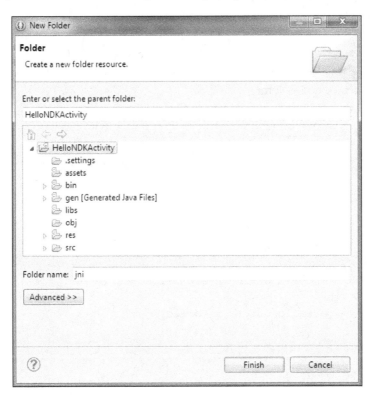

5.  Right-click on the newly-created `jni` folder under the `HelloNDK` project. Select **New | File**, enter `hello.c` as the value for **File name**, then click on **Finish**. Type the following code in the `hello.c` file:

```
#include <string.h>
#include <jni.h>

jstring
Java_cookbook_chapter1_HelloNDKActivity_naGetHelloNDKStr(JNIEnv*
pEnv, jobject pObj)
{
    return (*pEnv)->NewStringUTF(pEnv, "Hello NDK!");
}
```

6.  Right-click on the `jni` folder. Select **New | File**, enter `Android.mk` as the value for **File name**, then click on **Finish**. Type the following code in the `Android.mk` file:

```
LOCAL_PATH := $(call my-dir)
include $(CLEAR_VARS)
LOCAL_MODULE    := hello
LOCAL_SRC_FILES := hello.c
include $(BUILD_SHARED_LIBRARY)
```

7.  Start a terminal, go to the `jni` folder, and type `ndk-build` to build the `hello.c` program as a native library.

8.  Edit the `HelloNDKActivity.java` file. The file should contain the following content:

```
public class HelloNDKActivity extends Activity {
    @Override
    public void onCreate(Bundle savedInstanceState) {
        super.onCreate(savedInstanceState);
        TextView tv = new TextView(this);
        tv.setTextSize(30);
        tv.setText(naGetHelloNDKStr());
        this.setContentView(tv);
    }
    public native String naGetHelloNDKStr();
    static {
        System.loadLibrary("hello");
    }
}
```

9.  Right-click on the `HelloNDK` project in Eclipse. Select **Run As | Android Application**. Your Android phone or emulator will be displayed with something similar to the following screenshot:

## How it works...

This recipe demonstrated how to write a Hello NDK program on Android.

▶ **Native code**: The Hello NDK program consists of both the native C code and Java code. The native function `naGetHelloNDKStr` returns the `Hello NDK` string to the caller, as indicated in both the native code function definition and Java code method declaration. The native function name must follow a specific pattern for a package name, class name, and method name. The package and class name must agree with the package and class name of the Java class from which the native method is called, while the method name must be the same as the method name declared in that Java class.

This helps the Dalvik VM to locate the native function at runtime. Failing to follow the rule will result in `UnsatisfiedLinkError` at runtime.

The native function has two parameters, which are standard for all native functions. Additional parameters can be defined based on needs. The first parameter is a pointer to `JNIEnv`, which is the gateway to access various JNI functions. The meaning of the second parameter depends on whether the native method is a static or an instance method. If it's a static method, the second parameter is a reference to the class where the method is defined. If it's an instance method, the second parameter is a reference to the object on which the native method is invoked. We will discuss JNI in detail in *Chapter 2, Java Native Interface*.

▶ **Compilation of the native code**: The Android NDK build system frees developers from writing `makefile`. The build system accepts an `Android.mk` file, which simply describes the sources. It will parse the file to generate `makefile` and do all the heavy lifting for us.

We will cover details of how to write the `Android.mk` file or even write our own `makefile` in *Chapter 3, Build and Debug NDK Applications*.

Once we compile the native code, a folder named `libs` will be created under our project and a `libhello.so` library will be generated under the `armeabi` subdirectory.

- ▶ **Java code**: Three steps are followed to call the native method:

  1. **Load the native library**: This is done by calling `System.loadLibrary("hello")`. Note that instead of `libhello`, we should use `hello`. The Dalvik VM will fail to locate the library if `libhello` is specified.

  2. **Declare the method**: We declare the method with a native keyword to indicate that it is a native method.

  3. **Invoke the method**: We call the method just like any normal Java method.

## There's more...

The name of a native method is lengthy and writing it manually is error-prone. Fortunately, the `javah` program from JDK can help us generate the header file, which includes the method name. The following steps should be followed to use `javah`:

1. Write the Java code, including the native method definition.

2. Compile the Java code and make sure the class file appears under the `bin/classes/` folder of our project.

3. Start a terminal and go to the `jni` folder, and enter the following command:

   ```
   $ javah -classpath ../bin/classes -o <output file name> <java
   package name>.<java class anme>
   ```

   In our `HelloNDK` example, the command should be as follows:

   ```
   $ javah -classpath ../bin/classes -o hello.h cookbook.chapter1.
   HelloNDKActivity
   ```

   This will generate a file named `hello.h` with its function definition as follows:

   ```
   JNIEXPORT jstring JNICALL Java_cookbook_chapter1_HelloNDKActivity_
   naGetHelloNDKStr
     (JNIEnv *, jobject);
   ```

# 2
# Java Native Interface

In this chapter, we will cover the following recipes:

- ▶ Loading native libraries and registering native methods
- ▶ Passing parameters and receiving returns in primitive types
- ▶ Manipulating strings in JNI
- ▶ Managing references in JNI
- ▶ Manipulating classes in JNI
- ▶ Manipulating objects in JNI
- ▶ Manipulating arrays in JNI
- ▶ Accessing Java static and instance fields in native code
- ▶ Calling static and instance methods from native code
- ▶ Caching jfieldID, jmethodID, and reference data to improve performance
- ▶ Checking errors and handling exceptions in JNI
- ▶ Integrating assembly code in JNI

# Introduction

Programming with Android NDK is essentially writing code in both Java and native languages such as C, C++, and assembly. Java code runs on Dalvik **Virtual Machine** (**VM**), while native code is compiled to binaries running directly on the operating system. **Java Native Interface** (**JNI**) acts like the bridge that brings both worlds together. This relationship between Java code, Dalvik VM, native code, and the Android system can be illustrated using the following diagram:

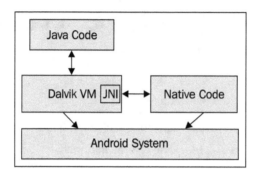

The arrow in the diagram indicates which party initiates the interaction. Both **Dalvik VM** and **Native Code** run on top of **Android system** (Android is a Linux-based OS). They require the system to provide the execution environment. **JNI** is part of **Dalvik VM**, which allows **Native Code** to access fields and invoke methods at Java Code. **JNI** also allows **Java Code** to invoke native methods implemented in **Native Code**. Therefore, **JNI** facilitates the two-way communication between **Native Code** and **Java Code**.

If you are familiar with Java programming and C, or C++, or assembly programming, then learning programming with Android NDK is mostly learning JNI. JNI comes with both primitive and reference data types. These data types have their corresponding mapping data types in Java. Manipulating the primitive types can usually be done directly, since a data type is normally equivalent to a native C/C++ data type. However, reference data manipulation often requires the help of the predefined JNI functions.

In this chapter, we'll first cover various data types in JNI and demonstrate how to invoke native methods from Java. We then describe accessing the Java fields and calling Java methods from the native code. Finally, we will discuss how to cache data to achieve better performance, how to handle errors and exceptions, and how to use assembly in native method implementation.

Every recipe in this chapter comes with a sample Android project that illustrates the topic and related JNI functions. Because of the space constraint, we cannot list all the source code in the book. The code is a very important part of this chapter and it is strongly recommended that you download the source code and refer to it when going through the recipes.

JNI is a complex topic, and we tried to cover the most essential parts of it in the context of Android NDK programming. However, a single chapter is not enough to provide all the details. Readers may want to refer to Java JNI Specification at `http://docs.oracle.com/javase/6/docs/technotes/guides/jni/` or the Java Native Interface: Programmer's Guide and Specification book at `http://java.sun.com/docs/books/jni/`. For Android-specific information, you can refer to JNI Tips at `https://developer.android.com/guide/practices/jni.html`.

# Loading native libraries and registering native methods

Native code is usually compiled into a shared library and loaded before the native methods can be called. This recipe covers how to load native libraries and register native methods.

## Getting ready

Please read the recipes in *Chapter 1, Hello NDK*, to set up the Android NDK development environment if you haven't done so already.

## How to do it...

The following steps will show you how to build an Android application that demonstrates loading native libraries and registering native methods:

1.  Start Eclipse, select **File | New | Android Project**. Enter the value for **Project Name** as `NativeMethodsRegister`. Select **Create new project in workspace**. Then, click on **Next**.

2.  In the next window, select the latest version of Android SDK, then click on **Next** to go to the next window.

3.  Specify the package name as `cookbook.chapter2`. Select the **Create Activity** checkbox, and specify the name as `NativeMethodsRegisterActivity`. Set the value for **Minimum SDK** as **5 (Android 2.0)**. Then, click on **Finish**.

4.  In **Eclipse Package Explorer**, right-click on the `NativeMethodsRegister` project, then select **New | Folder**. Enter the name `jni` in the pop-up window, then click on **Finish**.

5.  Right-click on the newly created `jni` folder under the `NativeMethodsRegister` project, then select **New | File**. Enter `nativetest.c` as the value for **File name**, then click on **Finish**.

6. Add the following code to `nativetest.c`:

```c
#include <android/log.h>
#include <stdio.h>

jint NativeAddition(JNIEnv *pEnv, jobject pObj, jint pa, jint pb)
{
  return pa+pb;
}


jint NativeMultiplication(JNIEnv *pEnv, jobject pObj, jint pa,
jint pb) {
  return pa*pb;
}


JNIEXPORT jint JNICALL JNI_OnLoad(JavaVM* pVm, void* reserved)
{
    JNIEnv* env;
    if ((*pVm)->GetEnv(pVm, (void **)&env, JNI_VERSION_1_6)) {
     return -1;
   }
    JNINativeMethod nm[2];
    nm[0].name = "NativeAddition";
    nm[0].signature = "(II)I";
    nm[0].fnPtr = NativeAddition;
    nm[1].name = "NativeMultiplication";
    nm[1].signature = "(II)I";
    nm[1].fnPtr = NativeMultiplication;
    jclass cls = (*env)->FindClass(env, "cookbook/chapter2/
NativeMethodRegisterActivity");
    // Register methods with env->RegisterNatives.
    (*env)->RegisterNatives(env, cls, nm, 2);
    return JNI_VERSION_1_6;
}
```

7. Add the following code to load the native shared library and define native methods to `NativeMethodRegisterActivity.java`:

```java
public class NativeMethodRegisterActivity extends Activity {
    ... ...
        private void callNativeMethods() {
          int a = 10, b = 100;
            int c = NativeAddition(a, b);
            tv.setText(a + "+" + b + "=" + c);
            c = NativeMultiplication(a, b);
            tv.append("\n" + a + "x" + b + "=" + c);
```

```
        }
        private native int NativeAddition(int a, int b);
        private native int NativeMultiplication(int a, int b);
        static {
           //use either of the two methods below
//System.loadLibrary("NativeRegister");
             System.load("/data/data/cookbook.chapter2/lib/
   libNativeRegister.so");
        }
   }
```

8.  Change `TextView` in the `res/layout/activity_native_method_register.xml` file as follows:

```
<TextView
        android:id="@+id/display_res"
        android:layout_width="wrap_content"
        android:layout_height="wrap_content"
        android:layout_centerHorizontal="true"
        android:padding="@dimen/padding_medium"
        android:text="@string/hello_world"
        tools:context=".NativeMethodRegisterActivity" />
```

9.  Create a file named `Android.mk` under the `jni` folder with the following content:

```
LOCAL_PATH := $(call my-dir)
include $(CLEAR_VARS)
LOCAL_MODULE     := NativeRegister
LOCAL_SRC_FILES := nativetest.c
LOCAL_LDLIBS := -llog
include $(BUILD_SHARED_LIBRARY)
```

10. Start a terminal, go to the `jni` folder under our project, and type `ndk-build` to build the native library.

11. Run the project on an Android device or emulator. You should see something similar to the following screenshot:

| NativeMethodRegisterActivity |
| --- |
| 10+100=110<br>10x100=1000 |

## How it works...

This recipe describes how to load a native library and register native methods:

▸ **Loading Native Library**: The `java.lang.System` class provides two methods to load native libraries, namely `loadLibrary` and `load`. `loadLibrary` accepts a library name without the prefix and file extension. For example, if we want to load the Android native library compiled as `libNativeRegister.so` in our sample project, we use `System.loadLibrary("NativeRegister")`. The `System.load` method is different. It requires the full path of the native library. In our sample project, we can use `System.load("/data/data/cookbook.chapter2/lib/libNativeRegister.so")` to load the native library. The `System.load` method can be handy when we want to switch between different versions of a native library, since it allows us to specify the full library path.

We demonstrated the usage of both the methods in the static initializer of the `NativeMethodRegisterActivity.java` class. Note that only one method should be enabled when we build and run the sample application.

▸ **JNIEnv Interface Pointer**: Every native method defined in native code at JNI must accept two input parameters, the first one being a pointer to `JNIEnv`. The `JNIEnv` interface pointer is pointing to thread-local data, which in turn points to a JNI function table shared by all threads. This can be illustrated using the following diagram:

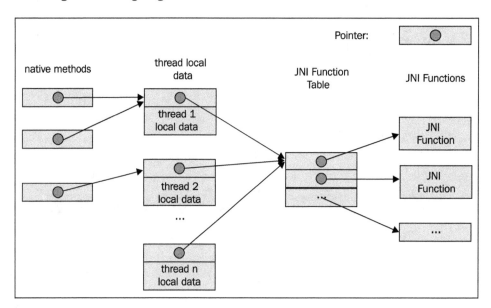

The JNIEnv interface pointer is the gateway to access all pre-defined JNI functions, including the functions that enable the native code to process Java objects, access Java fields, invoke Java methods, and so on. The RegisterNatives native function we're going to discuss next is also one of them.

The JNIEnv interface pointer points to thread-local data, so it cannot be shared between threads. In addition, JNIEnv is only accessible by a Java thread. A native thread must call the JNI function AttachCurrentThread to attach itself to the VM, to obtain the JNIEnv interface pointer. We will see an example of this in the *Manipulating classes in JNI* recipe in this chapter.

► **Registering Native Methods**: JNI can automatically discover the native method implementation if its function name follows a specific naming convention as mentioned in *Chapter 1, Hello NDK*. This is not the only way. In our sample project, we explicitly called the RegisterNatives JNI function to register the native methods. The RegisterNatives function has the following prototype:

```
jint RegisterNatives(JNIEnv *env, jclass clazz, const
JNINativeMethod *methods, jint nMethods);
```

The clazz argument is a reference to the class in which the native method is to be registered. The methods argument is an array of the JNINativeMethod data structure. JNINativeMethod is defined as follows:

```
typedef struct {
  char *name;
  char *signature;
  void *fnPtr;
} JNINativeMethod;
```

name indicates the native method name, signature is the descriptor of the method's input argument data type and return value data type, and fnPtr is the function pointer pointing to the native method. The last argument, nMethods of RegisterNatives, indicates the number of methods to register. The function returns zero to indicate success, and a negative value otherwise.

RegisterNatives is handy to register a native method implementation for different classes. In addition, it can simplify the native method name to avoid careless mistakes.

The typical way of using `RegisterNatives` is in the `JNI_OnLoad` method as shown in the following template. `JNI_OnLoad` is called when the native library is loaded, so we can guarantee that the native methods are registered before they're invoked:

```
JNIEXPORT jint JNICALL JNI_OnLoad(JavaVM* pVm, void* reserved)
{
    JNIEnv* env;
    if ((*pVm)->GetEnv(pVm, (void **)&env, JNI_VERSION_1_6)) {
    return -1;
    }

    // Get jclass with env->FindClass.
    // Register methods with env->RegisterNatives.

    return JNI_VERSION_1_6;
}
```

We demonstrated the usage of the preceding template in the `JNI_OnLoad` method of our sample code, where we registered two native methods to add and multiply two input integers respectively. The execution result shown earlier proves that the Java code can invoke the two registered native methods successfully.

Note that this example uses some JNI features that we're going to cover in later recipes, including the `FindClass` function and field descriptors. It is alright if don't fully understand the code at this stage. You can always go back to it after learning more about those topics.

# Passing parameters and receiving returns in primitive types

Java code can pass parameters to native methods and receive the processing results returned. This recipe walks through how to pass parameters and receive returns in primitive types.

## Getting ready

You should have built at least one Android application with native code before reading this recipe. If you haven't done so, please read the *Writing a Hello NDK program* recipe in *Chapter 1, Hello NDK* first.

## How to do it...

The following steps create a sample Android application with native methods receiving input parameters from the Java code and returning the processing result back:

1.  Create a project named `PassingPrimitive`. Set the package name as `cookbook.chapter2`. Create an activity named `PassingPrimitiveActivity`. Under this project, create a folder named `jni`. Please refer to the _Loading native libraries and registering native methods_ recipe in this chapter if you want more detailed instructions.

2.  Add a file named `primitive.c` under the `jni` folder and implement the native methods. In our sample project, we implemented one native method for each of the eight primitive data types. Following is the code for `jboolean`, `jint`, and `jdouble`. Please refer to the downloaded code for the complete list of methods:

```c
#include <jni.h>
#include <android/log.h>

JNIEXPORT jboolean JNICALL Java_cookbook_chapter2_
PassingPrimitiveActivity_passBooleanReturnBoolean(JNIEnv *pEnv,
jobject pObj, jboolean pBooleanP){
    __android_log_print(ANDROID_LOG_INFO, "native", "%d in %d
bytes", pBooleanP, sizeof(jboolean));
    return (!pBooleanP);
}


JNIEXPORT jint JNICALL Java_cookbook_chapter2_
PassingPrimitiveActivity_passIntReturnInt(JNIEnv *pEnv, jobject
pObj, jint pIntP) {
    __android_log_print(ANDROID_LOG_INFO, "native", "%d in %d
bytes", pIntP, sizeof(jint));
    return pIntP + 1;
}


JNIEXPORT jdouble JNICALL Java_cookbook_chapter2_
PassingPrimitiveActivity_passDoubleReturnDouble(JNIEnv *pEnv,
jobject pObj, jdouble pDoubleP) {
    __android_log_print(ANDROID_LOG_INFO, "native", "%f in %d
bytes", pDoubleP, sizeof(jdouble));
    return pDoubleP + 0.5;
}
```

3. In the `PassingPrimitiveActivity.java` Java code, we add code to load the native library, declare the native methods, and call the native methods. Following is that part of the code. The "..." indicates the part that is not shown. Please refer to the source file downloaded from the website for the complete code:

```
@Override
    public void onCreate(Bundle savedInstanceState) {
        super.onCreate(savedInstanceState);
        setContentView(R.layout.activity_passing_primitive);
        StringBuilder strBuilder = new StringBuilder();
        strBuilder.append("boolean: ").append(passBooleanReturnBoo
lean(false)).append(System.getProperty("line.separator"))
        . . . . . .

        .append("double: ").append(passDoubleReturnDoub
le(11.11)).append(System.getProperty("line.separator"));
        TextView tv = (TextView) findViewById(R.id.display_res);
        tv.setText(strBuilder.toString());
    }
    private native boolean passBooleanReturnBoolean(boolean p);
    private native byte passByteReturnByte(byte p);
    private native char passCharReturnChar(char p);
    private native short passShortReturnShort(short p);
    . . . . . .
    static {
        System.loadLibrary("PassingPrimitive");
    }
```

4. Modify the `res/layout/activity_passing_primitive.xml` file according to step 8 of the *Loading native libraries and registering native methods* recipe of this chapter or the downloaded project code.

5. Create a file named `Android.mk` under the `jni` folder, and add the following content to it:

```
LOCAL_PATH := $(call my-dir)
include $(CLEAR_VARS)
LOCAL_MODULE    := PassingPrimitive
LOCAL_SRC_FILES := primitive.c
LOCAL_LDLIBS := -llog
include $(BUILD_SHARED_LIBRARY)
```

6. Start a terminal, go to the `jni` folder, and type `ndk-build` to build the native library `PassingPrimitive`.

7. In Eclipse, select **Window** | **Show View** | **LogCat** to show the logcat console. Alternatively, start a terminal and enter the following command in your terminal to show `logcat` output on it:

```
$adb logcat -v time
```

8. Run the project on an Android device or emulator. You should see something similar to the following screenshot:

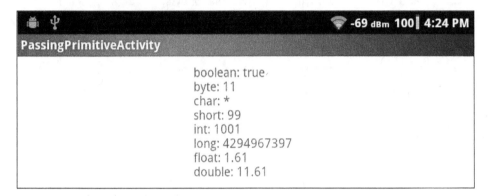

The logcat output is as follows:

```
6857    cookbook.chapter2    native    0 in 1 bytes
6857    cookbook.chapter2    native    10 in 1 bytes
6857    cookbook.chapter2    native    # in 2 bytes
6857    cookbook.chapter2    native    100 in 2 bytes
6857    cookbook.chapter2    native    1000 in 4 bytes
6857    cookbook.chapter2    native    4294967396 in 8 bytes
6857    cookbook.chapter2    native    1.110000 in 4 bytes
6857    cookbook.chapter2    native    11.110000 in 8 bytes
```

## How it works...

The code illustrates how to pass parameters and receive returns in primitive types from the native method. We created one method for each primitive type. In the native code, we printed the received value to `logcat`, modified the value, and returned it back.

- ▸ **JNI primitive type and Java primitive type mapping**: The primitive types in JNI and Java have the following mapping:

| Java Type | JNI Type | Number of bytes | Sign |
|-----------|----------|-----------------|------|
| boolean | jboolean | 1 | unsigned |
| byte | jbyte | 1 | signed |
| char | jchar | 2 | unsigned |
| short | jshort | 2 | signed |
| int | jint | 4 | signed |
| long | jlong | 8 | signed |
| float | jfloat | 4 | - |
| double | jdouble | 8 | - |

   Note that both Java `char` and JNI `jchar` are two bytes, while the C/C++ `char` type is only one byte long. In fact, C/C++ `char` are interchangeable with `jbyte` instead of `jchar` in JNI programming.

- ▸ **Android log library**: We output the received values to the Android logging system at a native method, by using the following code:

   ```
   __android_log_print(ANDROID_LOG_INFO, LOG_TAG, __VA_ARGS__);
   ```

   `ANDROID_LOG_INFO` is an `enum` value defined in `android/log.h`, which indicates that we're using the info-level logging. `LOG_TAG` can be any strings, and `__VA_ARGS__` is replaced by the parameters passed to the API, in a format similar to the `printf` method in C.

   We must include the `android/log.h` header in the native code to use the log functions:

   ```
   #include <android/log.h>
   ```

   Besides this, we'll need to include the NDK log library in the `Android.mk` file in order to use the API:

   ```
   LOCAL_LDLIBS := -llog
   ```

We will cover more details about Android logging API in *Chapter 3, Build and Debug NDK Applications*, while utilizing logging API for debugging purposes.

# Manipulating strings in JNI

Strings are somewhat complicated in JNI, mainly because Java strings and C strings are internally different. This recipe will cover the most commonly used JNI string features.

## Getting ready

Understanding the basics of encoding is essential to comprehend the differences between Java string and C string. We'll give a brief introduction to Unicode.

According to the Unicode Consortium, the Unicode Standard is defined as follows:

> *The Unicode Standard is a character coding system designed to support the worldwide interchange, processing, and display of the written texts of the diverse languages and technical disciplines of the modern world. In addition, it supports classical and historical texts of many written languages.*

Unicode assigns a unique number for each character it defines, called **code point**. There are mainly two categories of encoding methods that support the entire Unicode character set, or a subset of it.

The first one is the **Unicode Transformation Format** (**UTF**), which encodes a Unicode code point into a variable number of code values. UTF-8, UTF-16, UTF-32, and a few others belong to this category. The numbers 8, 16, and 32 refer to the number of bits in one code value. The second category is the **Universal Character Set** (**UCS**) encodings, which encodes a Unicode code point into a single code value. UCS2 and UCS4 belong to this category. The numbers 2 and 4 refer to the number of bytes in one code value.

Unicode defines more characters than what two bytes can possibly represent. Therefore, UCS2 can only represent a subset of Unicode characters. Because Unicode defines fewer characters than what four bytes can represent, multiple code values of UTF-32 are never needed. Therefore, UTF-32 and UCS4 are functionally identical.

Java programming language uses UTF-16 to represent strings. If a character cannot fit in a 16-bit code value, a pair of code values named **surrogate pair** is used. C strings are simply an array of bytes terminated by a null character. The actual encoding/decoding is pretty much left to the developer and the underlying system. A modified version of UTF-8 is used by JNI to represent strings, including class, field, and method names in the native code. There are two differences between the modified UTF-8 and standard UTF-8. Firstly, the null character is encoded using two bytes. Secondly, only one-byte, two-byte, and three-byte formats of Standard UTF-8 are supported by JNI, while longer formats cannot be recognized properly. JNI uses its own format to represent Unicode that cannot fit into three bytes.

## How to do it

The following steps show you how to create a sample Android project that illustrates string manipulation at JNI:

1.  Create a project named `StringManipulation`. Set the package name as `cookbook.chapter2`. Create an activity named `StringManipulationActivity`. Under the project, create a folder named `jni`. Refer to the *Loading native libraries and registering native methods* recipe in this chapter if you want more detailed instructions.

2.  Create a file named `stringtest.c` under the `jni` folder, then implement the `passStringReturnString` method as follows:

```c
JNIEXPORT jstring JNICALL Java_cookbook_chapter2_
StringManipulationActivity_passStringReturnString(JNIEnv *pEnv,
jobject pObj, jstring pStringP){

    __android_log_print(ANDROID_LOG_INFO, "native", "print
jstring: %s", pStringP);
  const jbyte *str;
  jboolean *isCopy;
  str = (*pEnv)->GetStringUTFChars(pEnv, pStringP, isCopy);
  __android_log_print(ANDROID_LOG_INFO, "native", "print UTF-8
string: %s, %d", str, isCopy);

    jsize length = (*pEnv)->GetStringUTFLength(pEnv, pStringP);
  __android_log_print(ANDROID_LOG_INFO, "native", "UTF-8 string
length (number of bytes): %d == %d", length, strlen(str));
  __android_log_print(ANDROID_LOG_INFO, "native", "UTF-8 string
ends with: %d %d", str[length], str[length+1]);
  (*pEnv)->ReleaseStringUTFChars(pEnv, pStringP, str);

  char nativeStr[100];
  (*pEnv)->GetStringUTFRegion(pEnv, pStringP, 0, length,
nativeStr);
  __android_log_print(ANDROID_LOG_INFO, "native", "jstring
converted to UTF-8 string and copied to native buffer: %s",
nativeStr);

    const char* newStr = "hello 安卓";
  jstring ret = (*pEnv)->NewStringUTF(pEnv, newStr);
  jsize newStrLen = (*pEnv)->GetStringUTFLength(pEnv, ret);
  __android_log_print(ANDROID_LOG_INFO, "native", "UTF-8 string
with Chinese characters: %s, string length (number of bytes)
%d=%d", newStr, newStrLen, strlen(newStr));
  return ret;
}
```

3. In the `StringManipulationActivity.java` Java code, add the code to load a native library, declare a native method, and invoke a native method. Refer to downloaded code for the source code details.

4. Modify the `res/layout/activity_passing_primitive.xml` file according to step 8 of the *Loading native libraries and registering native methods* recipe in this chapter or the downloaded project code.

5. Create a file called `Android.mk` under the `jni` folder. Refer to step 9 of the *Loading native libraries and registering native methods* recipe in this chapter or the downloaded code for details.

6. Start a terminal, go to the `jni` folder, and type `ndk-build` to build the native library.

7. Run the project on an Android device or emulator. We should see something similar to the following screenshot:

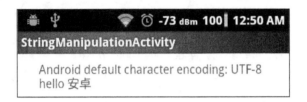

The following should be seen at the logcat output:

```
I  181 native print jstring: x
I  181 native print UTF-8 string: hello native code, 0
I  181 native UTF-8 string length (number of bytes): 17 == 17
I  181 native UTF-8 string ends with: 0 0
I  181 native jstring converted to UTF-8 string and copied to native buffer: hello native code
I  181 native UTF-8 string with Chinese characters: hello 安卓, string length (number of bytes) 12=12
```

## How it works...

This recipe discusses string manipulation at JNI.

► **Character encoding**: Android uses UTF-8 as its default charset, which is shown in our program by executing the `Charset.defaultCharset().name()` method. This means that the default encoding in the native code is UTF-8. As mentioned before, Java uses the UTF-16 charset. This infers that an encoding conversion is needed when we pass a string from Java to the native code and vice versa. Failing to do so will cause unwanted results. In our example, we tried printing `jstring` directly in the native code, but the result was some unrecognizable characters.

Fortunately, JNI comes with a few pre-defined functions that do the conversion.

▶ **Java string to native string**: When a native method is called with an input parameter of string type, the string received needs to be converted to the native string first. Two JNI functions can be used for different cases.

The first function is GetStringUTFChars, which has the following prototype:

```
const jbyte * GetStringUTFChars(JNIEnv *env, jstring string,
jboolean *isCopy);
```

This function converts the Java string into an array of UTF-8 characters. If a new copy of the Java string content is made, isCopy is set to true when the function returns; otherwise isCopy is set to false and the returned pointer points to the same characters as the original Java string.

It is not predictable whether the VM will return a new copy of the Java string. Therefore, we must be careful when converting a large string, as the possible memory allocation and copy may affect the performance and even cause "out of memory" issues. Also note that if isCopy is set to false, we cannot modify the returned UTF-8 native string, because it will modify the Java string content and break the immutability property of the Java string.

Once we've finished all the operations with the converted native string, we should call ReleaseStringUTFChars to inform the VM that we don't need to access the UTF-8 native string anymore. The function has the following prototype, with the second parameter being the Java string and the third parameter being the UTF-8 native string:

```
void ReleaseStringUTFChars(JNIEnv *env, jstring string, const char
*utf);
```

The second function for conversion is GetStringUTFRegion, with the following prototype:

```
void GetStringUTFRegion(JNIEnv *env, jstring str, jsize start,
jsize len, char *buf);
```

The start and len parameters indicate the start position of the Java UTF-16 string and number of UTF-16 characters for conversion. The buf argument points to the location to store the converted native UTF-8 char array.

Let's compare the two methods. The first method may or may not require allocation of new memory for the converted UTF-8 string depending on whether VM decides to make a new copy or not, whereas the second method made use of a pre-allocated buffer to store the converted content. In addition, the second method allows us to specify the position and length of the conversion source. Therefore, the following rules can be followed:

❑ To modify the converted UTF-8 native string, the JNI method GetStringUTFRegion should be used

- ❏ If we only need a substring of the original Java string, and the substring is not large, the `GetStringUTFRegion` should be used

- ❏ If we're dealing with a large string, and we're not going to modify the converted UTF-8 native string, `GetStringUTFChars` should be used

 In our example, we used a fixed length buffer when calling the `GetStringUTFRegion` function. We should make sure it is enough to hold the string, otherwise we should use the dynamic allocated array.

- ▶ **String length**: The JNI function `GetStringUTFLength` can be used to get the string length of a UTF-8 `jstring`. Note that it returns the number of bytes and not the number of UTF-8 characters, as shown in our example.

- ▶ **Native string to Java string**: We also need to return string data from the native code to Java code at times. The returned string should be UTF-16 encoded. The JNI function `NewStringUTF` constructs a `jstring` from a UTF-8 native string. It has the following prototype:

```
jstring NewStringUTF(JNIEnv *env, const char *bytes);
```

- ▶ **Conversion failure**: `GetStringUTFChars` and `NewStringUTF` require allocation of memory space to store the converted string. If you run out of memory, these methods will throw an `OutOfMemoryError` exception and return `NULL`. We'll cover more about exception handling in the *Checking errors and handling exceptions in JNI* recipe.

## There's more...

**More about character encoding in JNI**: JNI character encoding is much more complicated than what we covered here. Besides UTF-8, it also supports UTF-16 conversion functions. It is also possible to call Java string methods in the native code to encode/decode characters in other formats. Since Android uses UTF-8 as its platform charset, we only cover how to deal with conversions between Java UTF-16 and UTF-8 native string here.

# Managing references in JNI

JNI exposes strings, classes, instance objects, and arrays as reference types. The previous recipe introduces the string type. This recipe will cover reference management and the subsequent three recipes will discuss class, object, and arrays respectively.

## How to do it...

The following steps create a sample Android project that illustrates reference management in JNI:

1. Create a project named `ManagingReference`. Set the package name as `cookbook.chapter2`. Create an activity named `ManagingReferenceActivity`. Under the project, create a folder named `jni`. Refer to the *Loading native libraries and registering native methods* recipe in this chapter, if you want more detailed instructions.

2. Create a file named `referencetest.c` under the `jni` folder, then implement the `localReference`, `globalReference`, `weakReference`, and `referenceAssignmentAndNew` methods. This is shown in the following code snippet:

```c
JNIEXPORT void JNICALL Java_cookbook_chapter2_
ManagingReferenceActivity_localReference(JNIEnv *pEnv, jobject
pObj, jstring pStringP, jboolean pDelete){
    jstring stStr;
  int i;
  for (i = 0; i < 10000; ++i) {
    stStr = (*pEnv)->NewLocalRef(pEnv, pStringP);
    if (pDelete) {
      (*pEnv)->DeleteLocalRef(pEnv, stStr);
    }
  }
}


JNIEXPORT void JNICALL Java_cookbook_chapter2_
ManagingReferenceActivity_globalReference(JNIEnv *pEnv, jobject
pObj, jstring pStringP, jboolean pDelete){
  static jstring stStr;
  const jbyte *str;
  jboolean *isCopy;
  if (NULL == stStr) {
    stStr = (*pEnv)->NewGlobalRef(pEnv, pStringP);
  }
  str = (*pEnv)->GetStringUTFChars(pEnv, stStr, isCopy);
  if (pDelete) {
    (*pEnv)->DeleteGlobalRef(pEnv, stStr);
    stStr = NULL;
  }
}


JNIEXPORT void JNICALL Java_cookbook_chapter2_
ManagingReferenceActivity_weakReference(JNIEnv *pEnv, jobject
pObj, jstring pStringP, jboolean pDelete){
```

```
static jstring stStr;
const jbyte *str;
jboolean *isCopy;
if (NULL == stStr) {
  stStr = (*pEnv)->NewWeakGlobalRef(pEnv, pStringP);
}
str = (*pEnv)->GetStringUTFChars(pEnv, stStr, isCopy);
if (pDelete) {
  (*pEnv)->DeleteWeakGlobalRef(pEnv, stStr);
  stStr = NULL;
}
}
```

3. Modify the `ManagingReferenceActivity.java` file by adding code to load the native library, then declare and invoke the native methods.

4. Modify the `res/layout/activity_managing_reference.xml` file according to step 8 of the *Loading native libraries and registering native methods* recipe in this chapter, or the downloaded project code.

5. Create a file named `Android.mk` under the `jni` folder. Refer to step 9 of the *Loading native libraries and registering native methods* recipe of this chapter, or the downloaded code for details.

6. Start a terminal, go to the `jni` folder, and type `ndk-build` to build the native library.

7. Run the project on an Android device or emulator and monitor the logcat output with either eclipse or the `adb logcat -v time` command in your terminal. We'll show the sample results for each native method when while going through the details in the following section.

## How it works...

This recipe covers reference management in JNI:

▶ **JNI reference**: JNI exposes strings, classes, instance objects, and arrays as references. The basic idea of a reference can be illustrated using the following diagram:

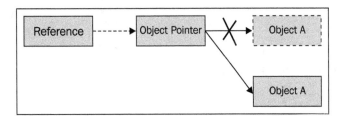

The reference adds one more level of indirection to an object (an object can be a class, an instance object, a string, or an array) access. An object is pointed by an object pointer, and a reference is used to locate the object pointer. Although this indirection introduces an overhead for object manipulation, it allows VM to conceal the object pointer from developers. The VM can therefore move the underlying object at runtime memory management and update the object pointer value accordingly, without affecting the reference.

Note that the garbage collector at VM moves the objects around to achieve cheap memory allocation, bulk de-allocation, reduce heap fragmentation, improve locality, and so on.

> A reference does not have to be a pointer. The details of how a reference is used to locate the object pointer are hidden from the developers.

▶ **Local reference versus global reference versus weak reference**: Three different types of references can be created to refer to the same data, namely local reference, global reference, and weak reference. Unless we explicitly create a global or weak reference, JNI operates using a local reference. The following table summarizes the differences between the three different types of references:

| | Creation | Lifespan | Visibility | Garbage collector (GC) behavior for referenced object | Free |
|---|---|---|---|---|---|
| **Local reference** | `Default` or `NewLocalRef` | One invocation of the native method. Invalid after native method returns. | Within the thread that created it. | GC won't garbage collect the referenced object. | Automatically freed or call `DeleteLocalRef` |
| **Global reference** | `NewGlobalRef` | Valid until freed explicitly. | Across multiple threads. | GC won't garbage collect the referenced object. | `DeleteGlobalRef` |
| **Weak reference** | `NewGlobalWeakRef` | Valid until freed explicitly. | Across multiple threads. | GC can garbage collect the referenced object. | `DeleteWeakGlobalRef` |

We will now take a look at the reference types one by one while referring to the sample source code:

- **Local reference**: The native method `localReference` shows the two basic JNI functions, namely `NewLocalRef` and `DeleteLocalRef`. The first function creates a local reference, while the second frees it. Note that normally we don't have to free a local reference explicitly, as it will be automatically freed after the native method returns. However, there are two exceptions. First, if we're creating lots of local references within a native method invocation, we can cause overflow. This is illustrated in our sample method when we pass `false` to the `pDelete` input parameter. The following screenshot represents an example of such a scenario:

```
I/localReference( 5840): delete after allocation: 1
I/localReference( 5840): finished
I/localReference( 5840): delete after allocation: 0
W/dalvikvm( 5840): ReferenceTable overflow (max=512)
W/dalvikvm( 5840): Last 10 entries in JNI local reference table:
W/dalvikvm( 5840):    502: 0x40510020 cls=Ljava/lang/String; (28 bytes)
W/dalvikvm( 5840):    503: 0x40510020 cls=Ljava/lang/String; (28 bytes)
```

The first execution deletes the local reference right after using it, so it's finished fine, while the second doesn't delete the local reference and eventually causes the `ReferenceTable` overflow.

Secondly, when we implement a utility function that is called by other native functions, we should not leak any references other than the return value. Otherwise, if the utility function is invoked by a native method many times, it will also cause an overflow issue.

Prior to Android 4.0, the local references were implemented using direct pointers to objects. Furthermore, those direct pointers were never invalidated even after `DeleteLocalRef` was called. Therefore, programmers can use local references as direct pointers, even after the reference is claimed to be deleted. A lot of JNI code not coded correctly worked due to this design. However, local references have been changed to use an indirection mechanism from Android 4.0 onwards. Hence, the buggy code using local references as direct pointers are going to break in Android 4.0 onwards. It is strongly recommended that you always follow JNI specifications.

- **Global reference**: The native method, `globalReference`, demonstrates a typical usage of a global reference. The global reference is retained when passing `false` to the `pDelete` input parameter, since it is a static variable. Next time the method is called, the static global reference will still reference to the same object. Therefore, we don't need to call `NewGlobalRef` again. This technique can save us from carrying out the same operation at every invocation of global reference.

  We invoke `globalReference` three times in the Java code, as follows:

  ```
  globalReference("hello global ref", false);
  globalReference("hello global ref 2", true);
  globalReference("hello global ref 3", true);
  ```

  The results should look similar to the following:

  ```
  I/globalReference( 5998): create global reference
  I/globalReference( 5998): print UTF-8 string: hello global ref
  I/globalReference( 5998): print UTF-8 string: hello global ref
  I/globalReference( 5998): global reference deleted
  I/globalReference( 5998): create global reference
  I/globalReference( 5998): print UTF-8 string: hello global ref 3
  I/globalReference( 5998): global reference deleted
  ```

  The string passed along with the first method call is retained, and therefore the first two invocations display the same string. After we delete the global reference at the end of the second call, the third call displays the string passed along with it.

  Note that although `DeleteGlobalRef` frees the global reference, it doesn't set it to `NULL`. We have explicitly set the global reference to `NULL` after the deletion.

- **Weak reference**: Weak reference is similar to global reference, except that it doesn't prevent the **Garbage Collector** (**GC**) from garbage collecting the underlying object referenced by it. Weak reference is not used as often as local and global reference. A typical use case is when we are referencing to lots of non-critical objects, and we don't want to prevent the GC from garbage collecting some of those objects when the GC thinks it's necessary.

   The support for weak references in Android is version dependent. Prior to Android 2.2, weak references were not implemented at all. Prior to Android 4.0, it can only be passed to `NewLocalRef`, `NewGlobalRef`, and `DeleteWeakGlobalRef`. From Android 4.0 onwards, Android has full support for weak references.

- **Assignment versus New<ReferenceType>Ref**: In the `referencetest.c` source code, we implemented the native `ReferenceAssignmentAndNew` method. This method illustrates the difference between assignment and allocating a new reference.

We passed the input jstring `pStringP` to the JNI function `NewGlobalRef` twice, to create two global references (`globalRefNew` and `globalRefNew2`), and assigned one of the global references to a variable `globalRefAssignment`. We then tested if they were all referencing to the same object.

Since `jobject` and `jstring` are actually pointers to void data type, we can print out their values as integers. Lastly, we invoked `DeleteGlobalRef` three times. The following is a screenshot of the Android logcat output:

```
I/ReferenceAssignmentAndNew( 5461): pStringP and globalRefNew: 1
I/ReferenceAssignmentAndNew( 5461): globalRefNew and globalRefNew2: 1
I/ReferenceAssignmentAndNew( 5461): globalRefAssignment and globalRefNew: 1
I/ReferenceAssignmentAndNew( 5461): pointer size: 4
I/ReferenceAssignmentAndNew( 5461): pStringP: 1079049504
I/ReferenceAssignmentAndNew( 5461): globalRefNew: 1079049504
I/ReferenceAssignmentAndNew( 5461): globalRefNew2: 1079049504
I/ReferenceAssignmentAndNew( 5461): globalRefAssignment: 1079049504
I/ReferenceAssignmentAndNew( 5461): call DeleteGlobalRef 1
I/ReferenceAssignmentAndNew( 5461): call DeleteGlobalRef 2
I/ReferenceAssignmentAndNew( 5461): call DeleteGlobalRef 3
W/dalvikvm( 5461): JNI: DeleteGlobalRef(0x4050fd20) failed to find entry (valid=1)
```

The first three lines indicate that the input jstring `pStringP`, two global references `globalRefNew` and `globalRefNew2`, and the assigned jstring `globalRefAssignment` all reference to the same object. Lines five to eight of the output show the same value, which means all the references themselves are equivalent. Lastly, the first two calls of `DeleteGlobalRef` succeed, while the last one fails.

The `New<ReferenceType>Ref` JNI function actually locates the underlying object and then adds a reference to the object. It allows multiple references added for the same object. Note that although our sample execution shows the values of references created by `New<ReferenceType>Ref` are the same, it is not guaranteed. It is possible that two object pointers pointing to the same object and references referencing to the same object are associated with the two different pointers.

It is recommended that you never rely on the value of a reference; you should use JNI functions instead. One example of this is to use `IsSameObject` and never use "==" to test if two references point to the same underlying object unless we test against `NULL`.

The number of `Delete<ReferenceType>Ref` calls must match the number of `New<ReferenceType>Ref` invocations. Fewer calls will potentially cause a memory leak, while having more calls will fail, as shown in the preceding result.

The assignment operation doesn't go through the VM, therefore it won't cause the VM to add a new reference.

Note that although we used a global reference for illustration, the principles also apply to local and weak references.

## There's more...

There's another method to manage the local references with JNI functions
`PushLocalFrame` and `PopLocalFrame`. Interested readers can refer to
the JNI specification for more information.

After attaching a native thread with `AttachCurrentThread`, the code running in the thread
would not free the local references until the thread is detached. The local reference should be
freed explicitly. In general, it is a good practice that we free local reference explicitly, as long
as we don't need it any more.

# Manipulating classes in JNI

The previous recipe discusses that Android JNI supports three different kinds of references.
The references are used to access the reference data types, including string, class, instance
object, and array. This recipe focuses on class manipulations in Android JNI.

## Getting ready

The *Managing References in NDK* recipe should be read first before going through this recipe.

## How to do it...

The following steps describe how to build a sample Android application that illustrates class
manipulation in JNI:

1. Create a project named `ClassManipulation`. Set the package name as
   `cookbook.chapter2`. Create an activity named `ClassManipulationActivity`.
   Under the project, create a folder named `jni`. Refer to the *Loading native
   libraries and registering native methods* recipe of this chapter if you want
   more detailed instructions.

2. Create a file named `classtest.c` under the `jni` folder, then implement
   the `findClassDemo`, `findClassDemo2`, `GetSuperclassDemo`, and
   `IsAssignableFromDemo` methods. We can refer to the downloaded
   `ClassManipulation` project source code.

3. Modify `ClassManipulationActivity.java` by adding code to load the native
   library, declare native methods, and invoke native methods.

4. Create a `Dummy` class and a `DummySubClass` subclass that extends the `Dummy`
   class. Create a `DummyInterface` interface and a `DummySubInterface`
   subinterface, which extends the `DummyInterface`.

5. Modify the `layout` XML file, add the `Android.mk` build file, and build the native library. Refer to steps 8 to 10 of the the *Loading native libraries and registering native methods* recipe of this chapter for details.

6. We're now ready to run the project. We'll present the output while discussing each native method in the following section.

## How it works...

This recipe demonstrates the manipulation of classes in JNI. We highlight a few points as follows:

▶ **Class descriptor**: A class descriptor refers to the name of a class or an interface. It can be derived by replacing the "`.`" character in Java with "`/`" in JNI programming. For example, the descriptor for class `java.lang.String` is `java/lang/String`.

▶ **FindClass and class loader**: The JNI function `FindClass` has the following prototype:

```
jclass FindClass(JNIEnv *env, const char *name);
```

It accepts a `JNIEnv` pointer and a class descriptor, and then locates a class loader to load the corresponding class. It returns a local reference to an initialized class, or `NULL` in case of failure. `FindClass` uses the class loader associated with the topmost method of the call stack. If it cannot find one, it will use the "system" class loader. One typical example is that after we create a thread and attach it to the VM, the topmost method of the call stack will be as follows:

```
dalvik.system.NativeStart.run(Native method)
```

This method is not part of our application code. Therefore the "system" class loader is used.

 A thread can be created at Java (called the managed thread or Java thread) or native code (called the native thread or non-VM thread). The native thread can be attached to a VM by calling the JNI function `AttachCurrentThread`. Once attached, the native thread works just like a Java thread, running inside a native method. It remains attached until the JNI function `DetachCurrentThread` is called.

In our `ClassManipulation` project, we illustrated `FindClass` with `findClassDemo` and `findClassDemo2` native methods. The `findClassDemo` method runs in a VM created thread. The `FindClass` call will locate the class loader properly. The `findClassDemo2` method creates a non-VM thread and attaches the thread to VM. It illustrates the case we described in the preceding section. The logcat output for calling the two native methods is as follows:

```
cookbook.chapter2    findClassDemo     find String class succeeded
cookbook.chapter2    findClassDemo     find Dummy class succeeded
cookbook.chapter2    pthread_dummy     after attach: find String class succeeded
cookbook.chapter2    pthread_dummy     after attach: find Dummy class failed
cookbook.chapter2    dalvikvm          threadid=9: thread exiting with uncaught exception (group=0x40015560)
cookbook.chapter2    AndroidRuntime    FATAL EXCEPTION: Thread-10
cookbook.chapter2    AndroidRuntime    java.lang.NoClassDefFoundError: [generic]
cookbook.chapter2    AndroidRuntime           at dalvik.system.NativeStart.main(Native Method)
```

As shown in the output, the non-VM thread loads the `String` class successfully but not the `Dummy` class defined by us. The way to work around this issue is to cache a reference to the `Dummy` class in the `JNI_OnLoad` method. We'll provide a detailed example in the *Caching jfieldID, jmethodID, and referencing data to improve performance* recipe.

► `GetSuperclass`: The JNI function `GetSuperclass` has the following prototype:

```
jclass GetSuperclass(JNIEnv *env, jclass clazz);
```

It helps us to find the superclass of a given class. If `clazz` is `java.lang.Object`, this function returns `NULL`; if it's an interface, it returns a local reference to `java.lang.Object`; if it's any other class, it returns a local reference to its superclass.

In our `ClassManipulation` project, we illustrated `GetSuperclass` with the `GetSuperclassDemo` native method. We created a `Dummy` class and a `DummyInterface` interface in Java code, where `DummySubClass` extends `Dummy`, and `DummySubInterface` extends `DummyInterface`. In the native method, we then invoked `GetSuperclass` to `java.lang.Object`, `DummySubClass`, and `DummySubInterface` respectively. The following is a screenshot of the logcat output:

```
cookbook.chapter2    GetSuperclassDem  superOfObject is NULL
cookbook.chapter2    GetSuperclassDem  superOfDummySubClass is not NULL
cookbook.chapter2    nativeGetName      class cookbook.chapter2.Dummy
cookbook.chapter2    GetSuperclassDem  superOfDummySubInterface is not NULL
cookbook.chapter2    nativeGetName      class java.lang.Object
```

As shown in the screenshot, `GetSuperclass` can find the superclass of `DummySubClass` successfully. In this native method, we used a utility function `nativeGetClassName`, where we called the `toString` method. We'll cover more about how to make such method calls in the *Calling instance and static methods in JNI* recipe.

▶ `IsAssignableFrom`: The JNI function `IsAssignableFrom` has the following prototype:

```
jboolean IsAssignableFrom(JNIEnv *env, jclass cls1, jclass cls2);
```

This function returns `JNI_TRUE` if `cls1` can be safely casted to `cls2`, and `JNI_FALSE` otherwise. We demonstrated its usage with the native method `IsAssignableFromDemo`. We obtained a local reference to `DummySubClass`, and called `GetSuperclass` to get a local reference to `Dummy`. Then, we called `IsAssignableFrom` to test if we can cast `DummySubClass` to `Dummy` and vice versa. The following is a screenshot of the logcat output:

```
cookbook.chapter2    IsAssignableFromDemo    From sub to super: 1
cookbook.chapter2    IsAssignableFromDemo    From super to sub: 0
```

As expected, the subclass can be safely cast to its superclass, but not the other way round.

 The JNI function `DefineClass` is not supported on Android. This is because the function requires the raw class data as input, and the Dalvik VM on Android doesn't use the Java bytecode or class files.

# Manipulating objects in JNI

The previous recipe shows how we can manipulate classes in Android JNI. This recipe describes how to manipulate instance objects in Android NDK programming.

## Getting ready

The following recipes should be read first before going through this recipe:

▶ *Managing references in JNI*

▶ *Manipulating classes in JNI*

## How to do it...

Now we'll create an Android project with native methods demonstrating the usage of JNI functions related to instance objects. Perform the following steps:

1. Create a project named `ObjectManipulation`. Set the package name as `cookbook.chapter2`. Create an activity named `ObjectManipulationActivity`. Under the project, create a folder named `jni`. Please refer to the *Loading native libraries and registering native methods* recipe in this chapter, if you want more detailed instructions.

2. Create a file named `objecttest.c` under the `jni` folder, then implement the `AllocObjectDemo`, `NewObjectDemo`, `NewObjectADemo`, `NewObjectVDemo`, `GetObjectClassDemo`, and `IsInstanceOfDemo` methods. You can refer to the downloaded `ObjectManipulation` project source code.

3. Modify `ObjectManipulationActivity.java` by adding code to load the native library, declare the native methods, and invoke them.

4. Create a `Dummy` class, and a `DummySub` class which extends `Dummy`. Create a `Contact` class with two fields `name` and `age`, a constructor, and a `getContactStr` method.

5. Modify the `layout` XML file, add the `Android.mk` build file, and build the native library. Refer to steps 8 to 10 of the *Loading native libraries and registering native methods* recipe of this chapter for more details.

6. We're now ready to run the project. We'll present the output while discussing each native method in the following section.

## How it works...

This recipe presents various methods for manipulating objects in JNI:

▶ **Create instance objects in the native code**: Four JNI functions can be used to create instance objects of a Java class in the native code, namely `AllocObject`, `NewObject`, `NewObjectA`, and `NewObjectV`. The `AllocObject` function creates an uninitialized object, while the other three methods take a constructor as an input parameter to create the object. The prototypes for the four functions are as follows:

```
jobject AllocObject(JNIEnv *env, jclass clazz);

jobject NewObject(JNIEnv *env, jclass clazz,jmethodID methodID,
...);

jobject NewObjectA(JNIEnv *env, jclass clazz,jmethodID methodID,
jvalue *args);
```

```
jobject NewObjectV(JNIEnv *env, jclass clazz,jmethodID methodID,
va_list args);
```

The `clazz` argument is a reference to the Java class of which we want to create an instance object. It cannot be an array class, which has its own set of JNI functions. `methodID` is the constructor method ID, which can be obtained using the `GetMethodID` JNI function.

For `NewObject`, a variable number of arguments can be passed after `methodID`, and the function will pass them to the constructor to create the instance object. `NewObjectA` accepts an array of type `jvalue`, and passses it to the constructor. `jvalue` is a union type and is defined as follows:

```
typedef union jvalue {
    jboolean z;
    jbyte    b;
    jchar    c;
    jshort   s;
    jint     i;
    jlong    j;
    jfloat   f;
    jdouble  d;
    jobject  l;
} jvalue;
```

`NewObjectV` passes an argument stored in `va_list` to the constructor. `va_list`, along with `va_start`, `va_end`, and `va_arg` enable us to access a variable number of input arguments for a function. The details are beyond the scope of this book. However, you can get a basic idea of how it works from the sample code provided.

In the Java code, we called all four native methods, each of which uses a different JNI function to create an instance object of the `Contact` class defined by us. We will then display the values of the name and age fields of all four `Contact` objects. The following is a screenshot of a sample run:

| ObjectManipulationActivity |
| --- |
| null:0<br>B:20<br>C:30<br>D:40 |

As shown, the instance object created by `AllocObject` is not initialized and therefore all fields contain the default value assigned by Java, while the other three methods create objects with the initial value passed by us.

▶ GetObjectClass: This JNI function has the following prototype:

```
jclass GetObjectClass(JNIEnv *env, jobject obj);
```

It returns a local reference to the class of the instance object obj. The obj argument must not be NULL, otherwise it will cause the VM to crash.

In our GetObjectClassDemo native method implementation, we obtained a reference to the Contact class and then called AllocObject to create an uninitialized object instance. In the Java code, we display the fields of the created object instance as follows:

| ObjectManipulationActivity |
| --- |
| null:0 |

As expected, the field values for the uninitialized instance Contact object are the default values assigned by Java.

▶ IsInstanceOf: This JNI function call has the following prototype:

```
jboolean IsInstanceOf(JNIEnv *env, jobject obj, jclass clazz);
```

It determines if the instance object obj is an instance of class clazz. We illustrated the usage of this function in the IsInstanceOfDemo native method. The method creates a local reference to the Dummy class and a local reference to the DummySub class, which is a sub class of Dummy. It then creates two objects, one for each class. The code then calls IsInstanceOf with each of the object references against each of the class references, making four checks in total. We send the output to logcat. A sample execution of this method gives the following result :

```
I/IsInstanceOfDemo(12186): dummyObj, dummyCls: 1
I/IsInstanceOfDemo(12186): dummyObj, dummySubCls: 0
I/IsInstanceOfDemo(12186): dummySubObj, dummyCls: 1
I/IsInstanceOfDemo(12186): dummySubObj, dummySubCls: 1
```

As the result shows, the Dummy instance object is an instance of the Dummy class but not DummySub class, while the DummySub instance object is an instance of the Dummy class and the DummySub class.

# Manipulating arrays in JNI

JNI exposes strings, classes, instance objects, and arrays as reference types. This recipe will discuss arrays in JNI.

## Getting ready

You should make sure you've read the following recipes before going through this recipe:

▸ *Managing references in JNI*

▸ *Manipulating classes in JNI*

## How to do it...

In this section, we will create a sample Android project that demonstrates how to manipulate arrays in JNI.

1. Create a project named `ArrayManipulation`. Set the package name as `cookbook.chapter2`. Create an activity named `ArrayManipulationActivity`. Under the project, create a folder named `jni`. Refer to the *Loading native libraries and registering native methods* recipe of this chapter for more detailed instructions.

2. Create a file named `arraytest.c` under the `jni` folder, then implement the `GetArrayLengthDemo`, `NewObjectArrayDemo`, `NewIntArrayDemo`, `GetSetObjectArrayDemo`, `GetReleaseIntArrayDemo`, `GetSetIntArrayRegionDemo`, and `GetReleasePrimitiveArrayCriticalDemo` native methods.

3. Modify `ArrayManipulationActivity.java` by adding code to load the native library, declare the native methods, and invoke them.

4. Create a `Dummy` class with a single integer field named `value`.

5. Modify the layout XML file, add the `Android.mk` build file, and build the native library. Refer to steps 8 to 10 of the *Loading native libraries and registering native methods* recipe of this chapter for more details.

6. We're now ready to run the project. We'll present the output while discussing each native method in the following section.

## How it works...

Arrays are represented by `jarray` or its subtypes such as `jobjectArray` and `jbooleanArray`. Similar to `jstring`, they cannot be accessed directly by native code like C arrays do. JNI provides various functions for accessing arrays:

▶ **Create new arrays**: JNI provides `NewObjectArray` and `New<Type>Array` functions to create arrays for objects and primitive types. Their function prototypes are as follows:

```
jarray NewObjectArray(JNIEnv *env, jsize length, jclass
elementType, jobject initialElement);
<ArrayType> New<Type>Array(JNIEnv *env, jsize length);
```

We demonstrate the usage of NewObjectArray in the native method `NewObjectArrayDemo`, where we create 10 instances of the `Dummy` class. The `length` parameter of the function indicates the number of objects to create, `elementType` is a reference to the class, and `initialElement` is the initialization value that is going to be set for all the created object instances in the array. In the Java code, we implemented the `callNewObjectArrayDemo` method, which calls the `NewObjectArrayDemo` native method to create an array of 10 `Dummy` objects, all with the value field set to 5. The execution result should look similar to the following screenshot:

| ArrayManipulationActivity |
|---|
| 5 5 5 5 5 5 5 5 5 5 |

As expected, the `value` field of all the objects created by `NewObjectArray` is 5.

The usage of `New<Type>Array` is shown in the native method `NewIntArrayDemo`, where we create an array of 10 integers using the JNI function `NewIntArray`, and then assign a value to each of the integers. All eight primitive types (`jboolean`, `jbyte`, `jchar`, `jshort`, `jint`, `jlong`, `jfloat`, and `jdouble`) of JNI have a corresponding `New<Type>Array` function to create an array of its type. Note that `NewIntArrayDemo` calls the `GetIntArrayElements` and `ReleaseIntArrayElements` JNI functions, which we'll discuss later in this recipe. In the Java code, we implemented a `callNewIntArrayDemo` method to call `NewIntArrayDemo` and display the integer array elements on the screen. The execution of `callNewIntArrayDemo` gives the following result:

| ArrayManipulationActivity |
|---|
| 0 1 2 3 4 5 6 7 8 9 |

As shown in the screenshot, the integer arrays are assigned with values from 0 to 9.

▶ `GetArrayLength`: This native function has the following prototype:

```
jsize GetArrayLength(JNIEnv *env, jarray array);
```

It accepts a reference to `jarray` and returns its length. We demonstrated its usage in the native method `GetArrayLengthDemo`. In the Java code, we implemented the `callGetArrayLengthDemo` method, which creates three arrays, including a `double` array, a `Dummy` object array, and a two-dimensional array of integers. The method calls the `GetArrayLengthDemo` native method to find the lengths for the three arrays. We output the array length to logcat in the native method. The sample execution output should look similar to the following screesnhot:

```
cookbook.chapter2      GetArrayLengthDemo      length: 10
cookbook.chapter2      GetArrayLengthDemo      length: 20
cookbook.chapter2      GetArrayLengthDemo      length: 30
```

▶ **Access object arrays**: JNI provides two functions to access object arrays, namely `GetObjectArrayElement` and `SetObjectArrayElement`. As its name suggests, the first one retrieves a reference to an object element of an array, while the second one sets the element of an object array. The two functions have the following prototype:

```
jobject GetObjectArrayElement(JNIEnv *env, jobjectArray array,
jsize index);
void SetObjectArrayElement(JNIEnv *env, jobjectArray array, jsize
index, jobject value);
```

In the two functions, the argument `array` refers to the object array and `index` is the position of the element. While the `get` function returns a reference to the object element, the `set` function sets the element according to the `value` argument.

We illustrate the usage of the two functions in native method `GetSetObjectArrayDemo`. The method accepts an object array and an object. It replaces the object at index one with the object received and then returns the original object at index one. In the Java code, we call the `callGetSetObjectArrayDemo` method to pass an array of three `Dummy` objects with values of 0, 1, 2, and another `Dummy` object of value `100` to the native method. The execution result should look similar to the following screenshot:

**ArrayManipulationActivity**

```
0
100
2
returned dummy: 1
```

As shown, the object at index `1` is replaced by the object with value `100`, and the original object of value `1` is returned.

▶ **Access arrays of primitive types**: JNI provides three sets of functions to access arrays of primitive types. We demonstrate them separately using three different native methods, all using `jintarray` as an example. Arrays of other primitive types are similar to integers.

Firstly, if we want to create a separate copy of `jintarray` in a native buffer, or only access a small portion of a large array, `GetIntArrayRegion`/`SetIntArrayRegion` functions are the proper choices. These two functions have the following prototype:

```
void GetIntArrayRegion(JNIEnv *env, jintArray array, jsize start,
jsize len, jint* buf);
void SetIntArrayRegion(JNIEnv *env, jintArray array, jsize start,
jsize len, jint* buf);
```

The two functions accept the same set of input parameters. The argument `array` refers to the `jintArray` we operate on, `start` is the start element position, `len` indicates the number of elements to get or set, and `buf` is the native integer buffer. We show the usage of these two functions in a native method called `GetSetIntArrayRegionDemo`. The method accepts an input `jintArray`, copies three elements from index 1 to 3 of the array to a native buffer, multiplies their values by 2 at the native buffer, and copies the value back to index 0 to 2.

In the Java code, we implement the `callGetSetIntArrayRegionDemo` method to initialize an integer array, pass the array to a native method `GetSetIntArrayRegionDemo`, and display the before and after values of all the elements. You should see an output similar to the following screenshot:

**ArrayManipulationActivity**

Before native method call:
0 1 2 3 4
After native method call:
2 4 6 3 4

The initial values for the five elements were 0, 1, 2, 3, and 4. We copied three elements from index one (1, 2, 3) to the native buffer `buf`. We then multiplied the values at the native buffer by 2, which made the first three elements at the native buffer 2, 4, and 6. We copied these three values from the native buffer back to the integer array, starting at index 0. The final values for the three elements were therefore 2, 4, and 6, and the last two elements remained unchanged as 3 and 4.

Secondly, if we want to access a large array, then `GetIntArrayElements` and `ReleaseIntArrayElements` are the JNI functions for us. They have the following prototype:

```
jint *GetIntArrayElements(JNIEnv *env, jintArray array, jboolean
*isCopy);
void ReleaseIntArrayElements(JNIEnv *env, jintArray array, jint
*elems, jint mode);
```

`GetIntArrayElements` returns a pointer to the array elements, or `NULL` in case of a failure. The array input parameter refers to the array we want to access, and `isCopy` is set to `true` if a new copy is created after the function call finishes. The returned pointer is valid until `ReleaseIntArrayElements` is called.

`ReleaseIntArrayElements` informs the VM that we don't need access to the array elements any more. The input parameter `array` refers to the array we operate on, `elems` is the pointer returned by `GetIntArrayElements`, and `mode` indicates the release mode. When `isCopy` at `GetIntArrayElements` is set to `JNI_TRUE`, the changes we make through the returned pointer will be reflected on the `jintArray`, since we're operating on the same copy. When `isCopy` is set to `JNI_FALSE`, the `mode` parameter determines how the data release is done. Depending upon whether we want to copy values from the native buffer back to the original array, and whether we want to free the `elems` native buffer, the `mode` parameters can be `0`, `JNI_COMMIT`, or `JNI_ABORT`, as follows:

| Copy values back / Free native buffer | Yes | No |
|---|---|---|
| Yes | 0 | JNI_ABORT |
| No | JNI_COMMIT | - |

We illustrate the two JNI functions with the native method `GetReleaseIntArrayDemo`. The method accepts an input integer array, obtains a native pointer through `GetIntArrayElements`, multiplies each element by 2, and finally commits the changes back by `ReleaseIntArrayElements` with `mode` set to 0. In the Java code, we implement the `callGetReleaseIntArrayDemo` method to initialize the input array and invoke the `GetReleaseIntArrayDemo` native method. The following is a screenshot of the phone display after executing the `callGetReleaseIntArrayDemo` method:

| ArrayManipulationActivity |
|---|
| Before native method call: |
| 0  1  2  3  4 |
| After native method call: |
| 0  2  4  6  8 |

As expected, all integer elements in the original array are multiplied by 2.

The third set of JNI functions are `GetPrimitiveArrayCritical` and `ReleasePrimitiveArrayCritical`. The usage of these two functions is similar to that of `Get<Type>ArrayElements` and `Release<Type>ArrayElements`, except for one important difference—the code block between the `Get` and `Release` methods is a critical region. No other JNI functions or function calls causing the current thread to wait for another thread in the same VM shall be made. These two methods essentially increase the possibility of obtaining an uncopied version of the original primitive array, and therefore improve the performance. We demonstrate the usage of these functions in a native method `GetReleasePrimitiveArrayCriticalDemo` along with the Java method `callGetReleasePrimitiveArrayCriticalDemo`. The implementations are similar to the second set of functions calls, and the display result is the same.

# Accessing Java static and instance fields in the native code

We have demonstrated how to pass parameters of different types to native methods and return data back to Java. This is not the only way of sharing data between the native code and Java code. This recipe covers another method—accessing Java fields from the native code.

## Getting ready

We're going to cover how to access Java fields of different types, including primitive types, strings, instance objects, and arrays. The following recipes should be read first before reading this recipe:

- *Passing parameters and receiving returns in primitive types*
- *Manipulating strings in JNI*
- *Manipulating classes in JNI*
- *Manipulating objects in JNI*
- *Manipulating arrays in JNI*

Readers are also expected to be familiar with Java reflection API.

## How to do it...

Follow these steps to create a sample Android project that demonstrates how to access Java static and instance fields from the native code:

1. Create a project named `AccessingFields`. Set the package name as `cookbook.chapter2`. Create an activity named `AccessingFieldsActivity`. Under the project, create a folder named `jni`. Refer to the *Loading native libraries and registering native methods* recipe of this chapter for more detailed instructions.

2. Create a file named `accessfield.c` under the `jni` folder, then implement the `AccessStaticFieldDemo`, `AccessInstanceFieldDemo`, and `FieldReflectionDemo` native methods.

3. Modify `AccessingFieldsActivity.java` by adding code to load the native library, declare native methods, and invoke them. In addition, add four instance fields and four static fields.

4. Create a `Dummy` class with an integer instance field named `value` and an integer static field named `value2`.

5. Modify the layout XML file, add the `Android.mk` build file, and build the native library. Refer to steps 8 to 10 of the *Loading native libraries and registering native methods* recipe of this chapter for more details.

6. We're now ready to run the project. We'll present the output while discussing each native method, in the following section.

# How it works...

This recipe discusses the access of fields (both static and instance fields) in Java from native code:

- ▸ **jfieldID data type**: `jfieldID` is a regular C pointer pointing to a data structure with details hidden from developers. We should not confuse it with `jobject` or its subtypes. `jobject` is a reference type corresponding to `Object` in Java, while `jfieldID` doesn't have such a corresponding type in Java. However, JNI provides functions to convert the `java.lang.reflect.Field` instance to `jfieldID` and vice versa.

- ▸ **Field descriptor**: It refers to the modified UTF-8 string used to represent the field data type. The following table summarizes the Java field types and its corresponding field descriptors:

| Java field type | Field descriptor |
| --- | --- |
| boolean | Z |
| byte | B |
| char | C |
| short | S |
| int | I |
| long | J |
| float | F |
| double | D |
| String | Ljava/lang/String; |
| Object | Ljava/lang/Object; |
| int [] | [I |
| Dummy [] | [Lcookbook/chapter2/Dummy; |
| Dummy [] [] | [[Lcookbook/chapter2/Dummy; |

As shown in the table, each of the eight primitive types has a single character string as its field descriptor. For objects, the field descriptor starts with "`L`", followed by the class descriptor (refer to the *Manipulating classes in JNI* recipe for detailed information) and ends with "`;`". For arrays, the field descriptor starts with "`[`", followed by the descriptor for the element type.

- **Accessing static fields**: JNI provides three functions to access static fields of a Java class. They have the following prototypes:

```
jfieldID GetStaticFieldID(JNIEnv *env, jclass clazz, const char
*name, const char *sig);
<NativeType> GetStatic<Type>Field(JNIEnv *env,jclass clazz,
jfieldID fieldID);
void SetStatic<Type>Field(JNIEnv *env, jclass clazz, jfieldID
fieldID,<NativeType> value);
```

To access a static field, the first step is to obtain the field ID, which is done by the first function listed here. In the method prototype, the clazz argument refers to the Java class at which the static field is defined, name indicates the field name, and sig is the field descriptor.

Once we have the method ID, we can either get or set the field value by calling function two or three. In the function prototype, <Type> can refer to any of the eight Java primitive types or Object, and fieldID is jfieldID returned by calling the first method. For set functions, value is the new value that we want to assign to the field.

The usage of the preceding three JNI functions are demonstrated in the native method AccessStaticFieldDemo, where we set and get values for an integer field, a string field, an array field, and a Dummy object field. These four fields are defined in the Java class AccessingFieldsActivity. In native code, we output the get values to Android logcat, while in the Java code we display the value set by the native code to the phone screen. The following screenshot shows the logcat output:

```
AccessStaticFieldDemo    sintF: 111
AccessStaticFieldDemo    sstrF: static string
AccessStaticFieldDemo    sintArrF 0: 1
AccessStaticFieldDemo    sintArrF 1: 2
AccessStaticFieldDemo    sintArrF 2: 3
AccessStaticFieldDemo    sdummy is instance of dummyCls: 1
AccessStaticFieldDemo    sdummyF value: 333
```

The phone display will look similar to the following screenshot:

**AccessingFieldsActivity**

sintF: 123
sstrF: hello from native
sintArrF: 0 1 2 3 4
sdummyF: 100

As shown, the values we set at the Java code for the fields can be obtained by the native code; and the values set by the native method are reflected in the Java code.

▶ **Accessing instance field**: Accessing instance fields is similar to accessing static fields. JNI also provides the following three functions for us:

```
jfieldID GetFieldID(JNIEnv *env, jclass clazz, const char *name,
const char *sig);
<NativeType> Get<Type>Field(JNIEnv *env,jobject obj, jfieldID
fieldID);
void Set<Type>Field(JNIEnv *env, jobject obj, jfieldID fieldID,
<NativeType> value);
```

Again, we need to obtain the field ID first, before we can get and set the values for the field. Instead of passing the class reference to the get and set functions, we should pass the object reference.

The usage is shown in native method AccessInstanceFieldDemo. Again, we print the values of get in the native code to the logcat and display the modified field values on the phone screen. The following screenshot shows the logcat output:

```
AccessInstanceFieldDemo     intF: 222
AccessInstanceFieldDemo     strF: instance string
AccessInstanceFieldDemo     intArrF 0: 1
AccessInstanceFieldDemo     intArrF 1: 2
AccessInstanceFieldDemo     intArrF 2: 3
AccessInstanceFieldDemo     intArrF 3: 4
AccessInstanceFieldDemo     intArrF 4: 5
AccessInstanceFieldDemo     sdummy is instance of dummyCls: 1
AccessStaticFieldDemo       dummyF value: 666
```

The phone display will look similar to the following screenshot:

**AccessingFieldsActivity**

```
intF: 123
strF: hello from native 2
intArrF: 0  10  20
dummyF: 200
```

A similar interpretation to accessing static fields can be made on the results.

▶ **Reflection support for field**: JNI provides two functions to support the interoperation with the Java Reflection API for `Field`. They have the following prototypes:

```
jfieldID FromReflectedField(JNIEnv *env, jobject field);
jobject ToReflectedField(JNIEnv *env, jclass cls, jfieldID
fieldID, jboolean isStatic);
```

The first function converts `java.lang.reflect.Field` to `jfieldID`, and then we can use the `set` and `get` JNI functions described previously. The argument field is an instance of `java.lang.reflect.Field`.

The second function does the reverse. It accepts a class reference, a `jfieldID`, and a `jboolean` variable indicating whether it is a static or an instance field. The function returns a reference to an object of `java.lang.reflect.Field`.

The usage of these two functions is demonstrated in the native method `FieldReflectionDemo`. We used the `Field` instance passed from the caller to access the field value, and then returned a `Field` instance for another field. In the Java method `callFieldReflectionDemo`, we pass the `Field` instance to the native code and use the returned `Field` instance to obtain the `field` value. The native code outputs the field value to logcat as follows:

```
FieldReflectionDemo          dummy value: 333
```

The Java code displays the value for another field on the phone screen as follows:

| AccessingFieldsActivity |
|---|
| value2: 100 |

# Calling static and instance methods from the native code

The previous recipe covers how to access Java fields in NDK. Besides fields, a Java class also has methods. This recipe focuses on calling static and instance methods from JNI.

## Getting ready

The code examples require a basic understanding of the JNI primitive types, strings, classes, and instance objects. It is better to make sure you have read the following recipes before going through this recipe:

▶ *Passing parameters and receiving returns in primitive types*

▶ *Manipulating strings in JNI*

> ▸ *Manipulating classes in JNI*

> ▸ *Manipulating objects in JNI*

> ▸ *Accessing Java static and instance fields in native code*

Readers are also expected to be familiar with Java reflection API.

## How to do it...

The following steps can be followed to create a sample Android project that illustrates how to call static and instance methods from the native code:

1. Create a project named `CallingMethods`. Set the package name as `cookbook.chapter2`. Create an activity named `CallingMethodsActivity`. Under the project, create a folder named `jni`. Refer to the *Loading native libraries and registering native methods* recipe of this chapter for more detailed instructions.

2. Create a file named `callmethod.c` under the `jni` folder, then implement the native methods `AccessStaticMethodDemo`, `AccessInstanceMethodDemo`, and `MethodReflectionDemo`.

3. Modify `CallingMethodsActivity.java` by adding code to load the native library, declare the native methods, and invoke them.

4. Create a `Dummy` class with an integer instance field named `value` and an integer static field named `value2`. In addition, create a `DummySub` class that extends `Dummy` with an additional String field called `name`.

5. Modify the layout XML file, add the `Android.mk` build file, and build the native library. Refer to steps 8 to 10 of the *Loading native libraries and registering native methods* recipe of this chapter for more details.

6. We're now ready to run the project. We'll present the output while discussing each native method in the following section.

## How it works...

This recipe illustrates how to call the Java static and instance methods from the native code:

> ▸ **jmethodID data type**: Similar to `jfieldID`, `jmethodID` is a regular C pointer pointing to a data structure with details hidden from the developers. JNI provides functions to convert the `java.lang.reflect.Method` instance to `jmethodID` and vice versa.

- **Method descriptor**: This is a modified UTF-8 string used to represent the input (input arguments) data types and output (return type) data type of the method. Method descriptors are formed by grouping all field descriptors of its input arguments inside a "()", and appending the field descriptor of the return type. If the return type is void, we should use "V". If there's no input arguments, we should simply use "()", followed by the field descriptor of the return type. For constructors, "V" should be used to represent the return type. The following table lists a few Java methods and their corresponding method descriptors:

| Java method | Method descriptor |
| --- | --- |
| Dummy(int pValue) | (I)V |
| String getName() | ()Ljava/lang/String; |
| void setName(String pName) | (Ljava/lang/String;)V |
| lont f(byte[] bytes, Dummy dummy) | ([BLcookbook/chapter2/Dummy;)J |

- **Calling static methods**: JNI provides four sets of functions for native code to call Java methods. Their prototypes are as follows:

```
jmethodID GetStaticMethodID(JNIEnv *env, jclass clazz, const char
*name, const char *sig);


<NativeType> CallStatic<Type>Method(JNIEnv *env, jclass clazz,
jmethodID methodID, ...);


<NativeType> CallStatic<Type>MethodA(JNIEnv *env, jclass clazz,
jmethodID methodID, jvalue *args);


<NativeType> CallStatic<Type>MethodV(JNIEnv *env, jclass
clazz,jmethodID methodID, va_list args);
```

The first function gets the method ID. It accepts a reference clazz to the Java class, a method name in a modified UTF-8 string format, and a method descriptor sig. The other three sets of functions are used to call the static methods. <Type> can be any of the eight primitive types, Void, or Object. It indicates the return type of the method invoked. The methodID argument is the jmethodID returned by the GetStaticMethodID function. The arguments to the Java method are passed one by one in CallStatic<Type>Method, or put into an array of jvalue as CallStatic<Type>MethodA, or put into the va_list structure as CallStatic<Type>MethodV.

We illustrate the usage of all the four sets of JNI functions in a native method `AccessStaticMethodDemo`. This method gets the method IDs for the `getValue2` and `setValue2` static methods of the `Dummy` class, and invokes these two methods using three different ways to pass the arguments to the called Java method. In `CallingMethodsActivity.java`, we implement `callAccessStaticMethodDemo`, which initializes the `value2` static field to `100`, invokes the native method `AccessStaticMethodDemo`, and prints the final `value2` value on phone screen. The following screenshot shows the logcat output:

```
AccessStaticMethodDemo          value2: 100
AccessStaticMethodDemo          value2 after set to 123: 123
AccessStaticMethodDemo          value2 after set to 124: 124
AccessStaticMethodDemo          value2 after set to 125: 125
```

The output of screen looks similar to the following screenshot:

**AccessingMethodsActivity**

dummy: 125

As shown, the native method firstly got `value2` as `100`, and it then used three different JNI functions to call the `set` method to modify the value. Finally, the phone display indicated that the final modified value is reflected in Java code.

▶ *Calling instance methods*: Calling instance methods from the native code is similar to calling static methods. JNI also provides four sets of functions as follows:

```
jmethodID GetMethodID(JNIEnv *env, jclass clazz, const char *name,
const char *sig);

<NativeType> Call<Type>Method(JNIEnv *env, jobject obj, jmethodID
methodID, ...);

<NativeType> Call<Type>MethodA(JNIEnv *env, jobject obj, jmethodID
methodID, jvalue *args);

<NativeType> Call<Type>MethodV(JNIEnv *env, jobject obj, jmethodID
methodID, va_list args);
```

The usage of these four sets of functions is similar to that of the JNI functions for calling static methods, except that we need to pass a reference to the instance object instead of the class. In addition, JNI provides another three sets of functions for calling instance methods, as follows:

```
<NativeType> CallNonvirtual<Type>Method(JNIEnv *env, jobject obj,
jclass clazz, jmethodID methodID, ...);


<NativeType> CallNonvirtual<Type>MethodA(JNIEnv *env, jobject obj,
jclass clazz, jmethodID methodID, jvalue *args);


<NativeType> CallNonvirtual<Type>MethodV(JNIEnv *env, jobject obj,
jclass clazz, jmethodID methodID, va_list args);
```

These three sets of methods accept an extra argument `clazz` as compared to the three sets of functions earlier. The `clazz` argument can be a reference to the class that `obj` is instantiated from, or a superclass of `obj`. A typical use case is to call `GetMethodID` on a class to obtain a `jmethodID`. We have a reference to an object of the class's subclass, and then we can use the preceding functions to call the Java method associated by `jmethodID` with the object reference.

The usage of all the seven sets of functions is illustrated in a native method `AccessInstanceMethodDemo`. We used the first four sets of functions to call `getName` and `setName` methods of the `DummySub` class with an object of it. We then used `CallNonvirtual<Type>Method` to call the `getValue` and `setValue` methods, which are defined in the `Dummy` superclass. In `CallingMethodsActivity.java`, we implemented the `callAccessInstanceMethodDemo` method to invoke the `AccessInstanceMethodDemo` native method. The following screenshot shows the logcat output:

```
cookbook.chapter2    AccessInstanceMethodDemo    name: A
cookbook.chapter2    AccessInstanceMethodDemo    name after set to B: B
cookbook.chapter2    AccessInstanceMethodDemo    name after set to C: C
cookbook.chapter2    AccessInstanceMethodDemo    name after set to C: D
cookbook.chapter2    AccessInstanceMethodDemo    value: 1
cookbook.chapter2    AccessInstanceMethodDemo    value: 123
```

As the results show, the `getName`, `setName`, `getValue`, and `setValue` methods are executed successfully.

▶ **Reflection support for method**: Similar to fields, JNI also provides the following two functions to support reflection:

```
jmethodID FromReflectedMethod(JNIEnv *env, jobject method);

jobject ToReflectedMethod(JNIEnv *env, jclass cls, jmethodID
methodID, jboolean isStatic);
```

The first function accepts a reference to the `java.lang.reflect.Method` instance, and returns its corresponding `jmethodID`. The returned `jmethodID` value can then be used to call the associated Java method. The second function does the reverse. It accepts a reference to the Java class, `jmethodID`, and `jboolean` indicating whether it's a static method or not, and returns a reference to `java.lang.reflect.Method`. The return value can be used in the Java code to access the corresponding method.

We illustrate these two JNI functions in native method `MethodReflectionDemo`. In `CallingMethodsActivity.java`, we implement the `callMethodReflectionDemo` method to pass the `java.lang.reflect.Method` object of `getValue` to the native code, get the returned `setValue java.lang.reflect.Method` object, and invoke the `setValue` method with the returned object.

The native method outputs the return value of `getValue` method to logcat as follows:

```
MethodReflectionDemo          value: 10
```

The Java code displays the `getValue` method return values before and after invoking `setValue` on the phone screen as follows:

```
AccessingMethodsActivity

    before native call: 10
    after invoke method returned from native method: 1234
```

As expected, the native code can access the `getValue` method with the `Method` object passed from the Java code, and the Java code can call the `setValue` method with the `Method` object returned from the native method.

# Caching jfieldID, jmethodID, and referencing data to improve performance

This recipe covers caching in Android JNI, which can improve the performance of our native code.

## Getting ready

You should make sure you've read the following recipes before going through this recipe:

- *Accessing Java static and instance fields in native code*
- *Calling static and instance methods from native code*

## How to do it...

The following steps detail how to build a sample Android application that demonstrates caching in JNI:

1. Create a project named `Caching`. Set the package name as `cookbook.chapter2`. Create an activity named `CachingActivity`. Under the project, create a folder named `jni`. Refer to the *Loading native libraries and registering native methods* recipe of this chapter for more detailed instructions.

2. Create a file named `cachingtest.c` under the `jni` folder, then implement the `InitIDs`, `CachingFieldMethodIDDemo1`, `CachingFieldMethodIDDemo2`, and `CachingReferencesDemo` methods.

3. Modify the `CachingActivity.java` file by adding code to load the native library, then declare and invoke the native methods.

4. Modify the layout XML file, add the `Android.mk` build file, and build the native library. Refer to steps 8 to 10 of the *Loading native libraries and registering native methods* recipe of this chapter for details.

5. Run the project on an Android device or emulator and monitor the logcat output with either eclipse or the `adb logcat -v time` command in your terminal.

6. At the `onCreate` method of `CachingActivity.java`, enable the `callCachingFieldMethodIDDemo1` method and disable the other demo methods. Start the Android application, and you should be able to see the following at logcat:

```
CachingFieldMethodIDDemo1  getValue returned 100: value field 100
CachingFieldMethodIDDemo1  getValue returned 200: value field 200
CachingFieldMethodIDDemo1  getValue returned 300: value field 300
```

7. Enable `callCachingFieldMethodIDDemo2` at `CachingActivity.java` while disabling the other demo methods and `InitIDs` method (at the static initializer). Start the Android application, and you should be able to see the following at logcat:

```
CachingFieldMethodIDDemo2   fid is null, get it
CachingFieldMethodIDDemo2   mid is null, get it
CachingFieldMethodIDDemo2   getValue returned 100: value field 100
CachingFieldMethodIDDemo2   getValue returned 200: value field 200
CachingFieldMethodIDDemo2   getValue returned 300: value field 300
```

8. Enable `callCachingReferencesDemo` at `CachingActivity.java` while commenting out other demo methods. Start the Android application, and you should be able to see the following at logcat:

```
CachingReferencesDemo     cached string is null, cache it
CachingReferencesDemo     caching string
CachingReferencesDemo     --------------
CachingReferencesDemo     caching string
CachingReferencesDemo     deleted the global reference
CachingReferencesDemo     --------------
CachingReferencesDemo     cached string is null, cache it
CachingReferencesDemo     caching string
CachingReferencesDemo     deleted the global reference
```

## How it works...

This recipe discusses the usage of caching at JNI:

▶ **Caching field and method IDs**: Field and method IDs are internal pointers. They're required to access a Java field or making native to Java method calls. Obtaining the field or method ID requires calling pre-defined JNI functions, which do symbolic lookups according to the name and descriptor. The lookup process typically requires several string comparisons and is relatively expensive.

Once the field or method ID is obtained, accessing the field or making native to Java calls is relatively quick. Therefore, a good practice is to perform lookup only once and cache the field or method ID.

There are two approaches to cache field and method IDs. The first approach caches at the class initializer. In Java, we can have something similar to the following:

```
private native static void InitIDs();
static {
    System.loadLibrary(<native lib>);
    InitIDs();
}
```

The static initializer is guaranteed to be executed before any of the class's methods. Therefore, we can ensure that the IDs required by the native method are valid when they're invoked. The usage of this approach is demonstrated in the `InitIDs` and `CachingFieldMethodIDDemo1` native methods and `CachingActivity.java`.

The second approach caches the IDs at the point of usage. We store the field or method ID in a static variable, so that the ID is valid the next time the native method is invoked. The usage of this approach is demonstrated in the native methods `CachingFieldMethodIDDemo2` and `CachingActivity.java`.

On comparison of these two approaches, the first one is preferred. Firstly, the first it doesn't require a validity check for the IDs before using them, because the static initializer is always called first and the IDs are therefore always valid before the native methods are called. Secondly, if the class is unloaded, the cached IDs will be invalid. If the second approach is used, we'll need to ensure the class is not unloaded and loaded again. If the first approach is used, the static initializer is called automatically when the class is loaded again, so we never have to worry about the class being unloaded and loaded again.

▶ **Caching references**: JNI exposes classes, instance objects, strings, and arrays as references. We covered how to manage references in the *Managing references at JNI* recipe. Sometimes, caching references can also improve performance. Unlike field and method IDs, which are direct pointers, references are implemented using an indirect mechanism that is hidden from developers. Therefore, we need to rely on JNI functions to cache them.

In order to cache reference data, we need to make it a global reference or weak global reference. A **global reference** guarantees that the reference will be valid until it is explicitly deleted. While **weak global** reference allows the underlying object to be garbage collected. Therefore, we'll need to do a validity check before using it.

The native method `CachingReferencesDemo` demonstrates how to cache a string reference. Note that while `DeleteGlobalRef` makes the global reference invalid, it doesn't assign `NULL` to the reference. We'll need to do this manually.

# Checking errors and handling exceptions in JNI

JNI functions can fail because of system constraint (for example, lack of memory) or invalid arguments (for example, passing a native UTF-8 string when the function is expecting a UTF-16 string). This recipe discusses how to handle errors and exceptions in JNI programming.

## Getting ready

The following recipes should be read first before proceeding with this recipe:

- *Manipulating strings in JNI*
- *Managing references in JNI*
- *Accessing Java static and instance fields in native code*
- *Calling static and instance methods from native code*

## How to do it...

Follow these steps to create a sample Android project that illustrates errors and exception handling in JNI:

1. Create a project named `ExceptionHandling`. Set the package name as `cookbook.chapter2`. Create an activity named `ExceptionHandlingActivity`. Under the project, create a folder named `jni`. Refer to the *Loading native libraries and registering native methods* recipe of this chapter for more detailed instructions.

2. Create a file named `exceptiontest.c` under the jni folder, then implement the `ExceptionDemo` and `FatalErrorDemo` methods.

3. Modify the `ExceptionHandlingActivity.java` file by adding code to load the native library, then declare and invoke the native methods.

4. Modify the layout XML file, add the `Android.mk` build file, and build the native library. Refer to steps 8 to 10 of the the *Loading native libraries and registering native methods* recipe of this chapter for more details.

5. We're now ready to run the project. We'll present the output while discussing each native method, in the following section.

## How it works...

This recipe discusses error checking and exception handling at JNI:

▸ **Check for errors and exceptions**: Many JNI functions return a special value to indicate failure. For example, the FindClass function returns NULL to indicate it failed to load the class. Many other functions do not use the return value to signal failure; instead an exception is thrown.

 Besides JNI functions, the Java code invoked by native code can also throw exceptions. We should make sure we check for such cases in order to write robust native code.

For the first group of functions, we can simply check the return value to see if an error occurs. For the second group of functions, JNI defines two functions to check for exceptions, as follows:

```
jboolean ExceptionCheck(JNIEnv *env);
jthrowable ExceptionOccurred(JNIEnv *env);
```

The first function returns JNI_TRUE to indicate that an exception occurs, and JNI_FALSE otherwise. The second function returns a local reference to the exception. When the second function is used, an additional JNI function can be called to examine the details of the exception:

```
void ExceptionDescribe(JNIEnv *env);
```

The function prints the exception and a back trace of the stack to the logcat.

In the native method ExceptionDemo, we used both approaches to check for occurrence of exceptions and ExceptionDescribe to print out the exception details.

▸ **Handle errors and exceptions**: Exceptions at JNI are different from Java exceptions. At Java, when an error occurs, an exception object is created and handed to the runtime. The runtime then searches the call stack for an exception handler that can handle the exception. The search starts at the method where the exception occurred and proceeds in the reverse order in which the methods are called. When such a code block is found, the runtime handles the control to the exception handler. The normal control flow is therefore interrupted. In contrast, JNI exception doesn't change the control flow, and we'll need to explicitly check for exception and handle it properly.

There are generally two ways to handle an exception. The first approach is to free the resources allocated at JNI and return. This will leave the responsibility of handling the exception to the caller of the native method.

The second practice is to clear the exception and continue executing. This is done through the following JNI function call:

```
void ExceptionClear(JNIEnv *env);
```

In the native method `ExceptionDemo`, we used the second approach to clear `java.lang.NullPointerException`, and the first approach to return `java.lang.RuntimeException` to the caller, which is the Java method `callExceptionDemo` at `ExceptionHandlingActivity.java`.

When an exception is pending, not all the JNI functions can be called safely. The following functions are allowed when there are pending exceptions:

- `DeleteGlobalRef`
- `DeleteLocalRef`
- `DeleteWeakGlobalRef`
- `ExceptionCheck`
- `ExceptionClear`
- `ExceptionDescribe`
- `ExceptionOccurred`
- `MonitorExit`
- `PopLocalFrame`
- `PushLocalFrame`
- `Release<PrimitiveType>ArrayElements`
- `ReleasePrimitiveArrayCritical`
- `ReleaseStringChars`
- `ReleaseStringCritical`
- `ReleaseStringUTFChars`

They're basically exception check and handle functions, or functions that clear resources allocated at native code.

 Calling JNI functions other than the functions listed here can lead to unexpected results when an exception is pending. We should handle the pending exception properly and then proceed.

▶ **Throw exceptions in the native code**: JNI provides two functions to throw an exception from native code. They have the following prototypes:

```
jint Throw(JNIEnv *env, jthrowable obj);
jint ThrowNew(JNIEnv *env, jclass clazz, const char *message);
```

The first function accepts a reference to a `jthrowable` object and throws the exception, while the second function accepts a reference to an exception class. It will create an exception object of the `clazz` class with the message argument and throw it.

In the `ExceptionDemo` native method, we used the `ThrowNew` function to throw `java.lang.NullPointerException` and a `Throw` function to throw `java.lang.RuntimeException`.

The following logcat output indicates how the exceptions are checked, cleared, and thrown:

```
ExceptionCheck: 0
ExceptionOccurred returned NULL? : 1
ExceptionCheck after finding non-existing class: 1
ExceptionCheck after clear: 0
null pointer exception thrown using ThrowNew
there's pending exception, call ExceptionDescribe
java.lang.NullPointerException: throw null pointer exception
        at cookbook.chapter2.ExceptionHandlingActivity.ExceptionDemo(Native Method)
        at cookbook.chapter2.ExceptionHandlingActivity.callExceptionDemo(ExceptionHandlingActivity.
java:22)
        at cookbook.chapter2.ExceptionHandlingActivity.onCreate(ExceptionHandlingActivity.java:17)
        at android.app.Instrumentation.callActivityOnCreate(Instrumentation.java:1047)
        at android.app.ActivityThread.performLaunchActivity(ActivityThread.java:1722)
        at android.app.ActivityThread.handleLaunchActivity(ActivityThread.java:1784)
        at android.app.ActivityThread.access$1500(ActivityThread.java:123)
        at android.app.ActivityThread$H.handleMessage(ActivityThread.java:939)
        at android.os.Handler.dispatchMessage(Handler.java:99)
        at android.os.Looper.loop(Looper.java:130)
        at android.app.ActivityThread.main(ActivityThread.java:3835)
        at java.lang.reflect.Method.invokeNative(Native Method)
        at java.lang.reflect.Method.invoke(Method.java:507)
        at com.android.internal.os.ZygoteInit$MethodAndArgsCaller.run(ZygoteInit.java:847)
        at com.android.internal.os.ZygoteInit.main(ZygoteInit.java:605)
        at dalvik.system.NativeStart.main(Native Method)
ExceptionCheck after ExceptionDescribe: 1
ExceptionCheck after clear: 0
exception thrown using Throw
```

The last exception is not cleared at the native method. In the Java code, we catch the exception and display the message on the phone screen:

**ExceptionHandlingActivity**

throw runtime exception

> ▶ **Fatal error**: A special type of error is the fatal error, which is not recoverable. JNI defines a function `FatalError`, as follows, to raise a fatal error:

```
void FatalError(JNIEnv *env, const char *msg);
```

This function accepts a message and prints it to logcat. After that, the VM instance for the application is terminated. We demonstrated the usage of this function in the native method `FatalErrorDemo` and Java method `callFatalErrorDemo`. The following output is captured at logcat:

```
JNI posting fatal error: fatal error
VM aborting
```

Note that the code after the `FatalError` function is never executed, in neither the native nor Java code, because `FatalError` never returns, and the VM instance is terminated. On my Android device, this does not lead the Android application to crash, but causes the application to freeze instead.

## There's more...

C++ exception is currently not supported on Android JNI programming. In other words, the native C++ exceptions do not propagate to Java world through JNI. Therefore, we should handle C++ exceptions within C++ code. Alternatively, we can write a C wrapper to throw an exception or return an error code to Java.

# Integrating assembly code in JNI

Android NDK allows you to write assembly code at JNI programming. Assembly code is sometimes used to optimize the critical portion of code to achieve the best performance. This recipe does not intend to discuss how to program in assembly. It describes how to integrate assembly code in JNI programming instead.

## Getting ready

Read the *Passing parameters and receiving returns in primitive types* recipe before you continue.

## How to do it...

The following steps create a sample Android project that integrates the assembly code:

1.  Create a project named `AssemblyInJNI`. Set the package name as `cookbook.chapter2`. Create an activity named `AssemblyInJNIActivity`. Under the project, create a folder named `jni`. Refer to the *Loading native libraries and registering native methods* recipe of this chapter for more detailed instructions.

2.  Create a file named `assemblyinjni.c` under the `jni` folder, then implement the `InlineAssemblyAddDemo` method.

3.  Create a file named `tmp.c` under the `jni` folder, and implement the native method `AssemblyMultiplyDemo`. Compile the `tmp.c` code to an assembly source file named `AssemblyMultiplyDemo.s`, using the following command:

    ```
    $ $ANDROID_NDK/toolchains/arm-linux-androideabi-4.4.3/
    prebuilt/linux-x86/bin/arm-linux-androideabi-gcc -S tmp.c
    -o AssemblyMultiplyDemo.s --sysroot=$ANDROID_NDK/platforms/
    android-14/arch-arm/
    ```

4.  Modify the `AssemblyInJNIActivity.java` file by adding code to load the native library, then declare and invoke the native methods.

5.  Modify the layout XML file, add the `Android.mk` build file, and build the native library. Refer to steps 8 to 10 of the *Loading native libraries and registering native methods* recipe of this chapter for details.

6.  At `AssemblyInJNIActivity.java`, enable the `callInlineAssemblyAddDemo` native method and disable the `callAssemblyMultiplyDemo` method. Run the project on an Android device or emulator. The phone display should look similar to the following screenshot:

    | AssemblyInJNIActivity |
    | --- |
    | 5+10=15 |

7.  At `AssemblyInJNIActivity.java`, enable the `callAssemblyMultiplyDemo` native method and disable the `callInlineAssemblyAddDemo` method. Run the project on an Android device or emulator. The phone display should look similar to the following screenshot:

    | AssemblyInJNIActivity |
    | --- |
    | 5x10=50 |

## How it works...

This recipe demonstrates the usage of the assembly code to implement a native method:

▶ **Inline assembly at C code**: We can write inline assembly code for Android NDK development. This is illustrated in native method `InlineAssemblyAddDemo`.

▶ **Generating a separate assembly code**: One approach to write assembly code is to write the code in C or C++, and use a compiler to compile the code into assembly code. Then, we optimize based on the auto-generated assembly code. Since this recipe is not about writing code in an assembly language, we use the Android NDK cross compiler to generate a native method `AssemblyMultiplyDemo` and call it from the Java method `callAssemblyMultiplyDemo`.

We first write the native method `AssemblyMultiplyDemo` in `AssemblyMultiplyDemo.c`, then cross compile the code using the compiler with Android NDK, using the following:

```
$ $ANDROID_NDK/toolchains/arm-linux-androideabi-4.4.3/prebuilt/
linux-x86/bin/arm-linux-androideabi-gcc -S <c_file_name>.c -o
<output_file_name>.s --sysroot=$ANDROID_NDK/platforms/android-
<level>/arch-<arch>/
```

In the preceding command, `$ANDROID_NDK` is an environment variable pointing to the location of Android NDK. If you have followed the recipes in *Chapter 1, Hello NDK*, then this should have been configured correctly. Otherwise, you can replace it with the full path to your Android NDK location (for example, in my computer, the path is `/home/roman10/Desktop/android/android-ndk-r8`). `<level>` indicates the targeted Android version. In our case, we used `14`. `<arch>` indicates the architecture; we used `arm`. If we build an application for other architectures such as x86, then this should be `x86`. The `-S` option tells the cross compiler to compile the `<c_file_name>.c` file into an assembly code, but don't assemble or link it. The `-o` option tells the compiler to output the assembly code to a file `<output_file_name>.s`. If no such option appears, the compiler outputs to a file named `<c_file_name>.s`.

▶ **Compile the assembly code**: Compiling assembly code is just like compiling C/C++ source code. As shown in the `Android.mk` file, we simply list the assembly file as a source file as follows:

```
LOCAL_SRC_FILES := AssemblyMultiplyDemo.s assemblyinjni.c
```

# 3
# Build and Debug NDK Applications

In this chapter we will cover the following recipes:

- ▶ Building an Android NDK application at the command line
- ▶ Building an Android NDK application in Eclipse
- ▶ Building an Android NDK application for different ABIs
- ▶ Building an Android NDK applications for different CPU features
- ▶ Debugging an Android NDK application with logging messages
- ▶ Debugging an Android NDK application with CheckJNI
- ▶ Debugging an Android NDK application with NDK GDB
- ▶ Debugging an Android NDK application with CGDB
- ▶ Debugging an Android NDK application in Eclipse

## Introduction

We covered the environment set up in *Chapter 1*, *Hello NDK*, and JNI programming in *Chapter 2*, *Java Native Interface*. To build Android NDK applications, we'll also need to use the **build** and **debug** tools for Android NDK.

Android NDK comes with the `ndk-build` script to facilitate the easy build of any Android NDK application. This script hides the complications of invoking cross compilers, cross linkers, and so on, from developers. We'll start by introducing the usage of the `ndk-build` command.

A recent release of the **Android Development Tools** (**ADT**) plugin has enabled the building of Android NDK applications from Eclipse. We'll demonstrate how to use it.

We'll explore building NDK applications for different **Application Binary Interfaces (ABIs)** and making use of optional CPU features. This is essential to achieve best performance on different Android devices.

Besides build, we will also introduce various debugging tools and techniques for Android NDK applications. Starting with the simple but powerful logging technique, we'll show how to debug NDK applications from both the command line and the Eclipse IDE. The CheckJNI mode will also be introduced, which can help us capture JNI bugs.

# Building an Android NDK application at the command line

Though Eclipse is the recommended IDE for Android development, sometimes we want to build an Android application in the command line so that the process can be automated easily and become part of a continuous integration process. This recipe focuses on how to build an Android DNK application at the command line.

## Getting ready

Apache Ant is a tool mainly used for building Java applications. It accepts an XML file to describe the build, deploy and test processes, manage the processes, and to automatically keep a track of the dependencies.

We are going to use Apache Ant to build and deploy our sample project. If you don't have it installed yet, you can follow these commands to install it:

- If you're on Ubuntu Linux, use the following command:

   ```
   $ sudo apt-get install ant1.8
   ```

- If you're using a Mac, use the following command:

   ```
   $ sudo port install apache-ant
   ```

- If you're using Windows, you can download the winant installer from http://code.google.com/p/winant/downloads/list, and install it.

Readers are supposed to have the NDK development environment set up and read the *Writing a Hello NDK program* recipe in *Chapter 1, Hello NDK*, before going through this one.

## How to do it...

The following steps create and build a sample `HelloNDK` application:

1. Create the project. Start a command-line console and enter the following command:

   ```
   $ android create project \
   --target android-15 \
   --name HelloNDK \
   --path ~/Desktop/book-code/chapter3/HelloNDK \
   --activity HelloNDKActivity \
   --package cookbook.chapter3
   ```

    The `android` tool can be found under the `tools/` directory of the Android SDK folder. If you have followed *Chapter 1, Hello NDK*, to set up the SDK and NDK development with PATH configured properly, you can execute the `android` command directly from the command line. Otherwise, you will need to enter the relative or full path to the `android` program. This also applies to other SDK and NDK tools used in the book.

   The following is a screenshot of the command output:

   ```
   roman10@roman10-laptop:~$ android create project --target android-15 --name HelloNDK --path ~/Desktop/b
   ook-code/chapter3/HelloNDK --activity HelloNDKActivity --package cookbook.chapter3
   Created project directory: /home/roman10/Desktop/book-code/chapter3/HelloNDK
   Created directory /home/roman10/Desktop/book-code/chapter3/HelloNDK/src/cookbook/chapter3
   Added file /home/roman10/Desktop/book-code/chapter3/HelloNDK/src/cookbook/chapter3/HelloNDKActivity.jav
   a
   Created directory /home/roman10/Desktop/book-code/chapter3/HelloNDK/res
   Created directory /home/roman10/Desktop/book-code/chapter3/HelloNDK/bin
   Created directory /home/roman10/Desktop/book-code/chapter3/HelloNDK/libs
   Created directory /home/roman10/Desktop/book-code/chapter3/HelloNDK/res/values
   Added file /home/roman10/Desktop/book-code/chapter3/HelloNDK/res/values/strings.xml
   Created directory /home/roman10/Desktop/book-code/chapter3/HelloNDK/res/layout
   Added file /home/roman10/Desktop/book-code/chapter3/HelloNDK/res/layout/main.xml
   Created directory /home/roman10/Desktop/book-code/chapter3/HelloNDK/res/drawable-hdpi
   Created directory /home/roman10/Desktop/book-code/chapter3/HelloNDK/res/drawable-mdpi
   Created directory /home/roman10/Desktop/book-code/chapter3/HelloNDK/res/drawable-ldpi
   Added file /home/roman10/Desktop/book-code/chapter3/HelloNDK/AndroidManifest.xml
   Added file /home/roman10/Desktop/book-code/chapter3/HelloNDK/build.xml
   Added file /home/roman10/Desktop/book-code/chapter3/HelloNDK/proguard-project.txt
   roman10@roman10-laptop:~$
   ```

2. Go to the `HelloNDK` project folder and create a folder named `jni` by using the following command:

```
$ cd ~/Desktop/book-code/chapter3/HelloNDK
$ mkdir jni
```

3. Create a file named `hello.c` under the `jni` folder, and add the following content to it:

```c
#include <string.h>
#include <jni.h>

jstring Java_cookbook_chapter3_HelloNDKActivity_
naGetHelloNDKStr(JNIEnv* pEnv, jobject pObj)
{
    return (*pEnv)->NewStringUTF(pEnv, "Hello NDK!");
}
```

4. Create a file named `Android.mk` under the `jni` folder with the following content:

```
LOCAL_PATH := $(call my-dir)
include $(CLEAR_VARS)
LOCAL_MODULE    := hello
LOCAL_SRC_FILES := hello.c
include $(BUILD_SHARED_LIBRARY)
```

5. Build the native library using the following command:

```
$ ndk-build
```

6. Modify the `HelloNDKActivity.java` file to the following content:

```java
package cookbook.chapter3;
import android.app.Activity;
import android.os.Bundle;
import android.widget.TextView;
public class HelloNDKActivity extends Activity {
    @Override
    public void onCreate(Bundle savedInstanceState) {
        super.onCreate(savedInstanceState);
        TextView tv = new TextView(this);
        tv.setTextSize(30);
        tv.setText(naGetHelloNDKStr());
        this.setContentView(tv);
    }
    public native String naGetHelloNDKStr();
    static {
        System.loadLibrary("hello");
    }
}
```

7.  Update the project. We have added a native library, so we need to update the project with the following command. Note that this command is only needed once unless we change the project settings, while the previous `ndk-build` command needs to be executed every time we update the native code:

    ```
    $ android update project --target android-15 --name HelloNDK \
    --path ~/Desktop/book-code/chapter3/HelloNDK
    ```

    The following is a screenshot of the command output:

    ```
    roman10@roman10-laptop:~/Desktop/book-code/chapter3/HelloNDK$ android update project --target android-1
    5 --name HelloNDK --path ~/Desktop/book-code/chapter3/HelloNDK
    Updated project.properties
    Updated local.properties
    Updated file /home/roman10/Desktop/book-code/chapter3/HelloNDK/build.xml
    Updated file /home/roman10/Desktop/book-code/chapter3/HelloNDK/proguard-project.txt
    It seems that there are sub-projects. If you want to update them
    please use the --subprojects parameter.
    roman10@roman10-laptop:~/Desktop/book-code/chapter3/HelloNDK$ []
    ```

8.  Go to the project `root` folder, and build our project in the debug mode using the following command:

    ```
    $ ant debug
    ```

    In the following screenshot, we show the last few lines of the output, which indicates a successful build is:

    ```
    -do-debug:
     [zipalign] Running zip align on final apk...
        [echo] Debug Package: /home/roman10/Desktop/book-code/chapter3/HelloNDK/bin/HelloNDK-debug.apk
    [propertyfile] Creating new property file: /home/roman10/Desktop/book-code/chapter3/HelloNDK/bin/build.
    prop
    [propertyfile] Updating property file: /home/roman10/Desktop/book-code/chapter3/HelloNDK/bin/build.prop
    [propertyfile] Updating property file: /home/roman10/Desktop/book-code/chapter3/HelloNDK/bin/build.prop
    [propertyfile] Updating property file: /home/roman10/Desktop/book-code/chapter3/HelloNDK/bin/build.prop

    -post-build:

    debug:

    BUILD SUCCESSFUL
    Total time: 3 seconds
    ```

    The output `apk` will be produced at `bin/HelloNDK-debug.apk`.

9.  Create an emulator using the following command:

    ```
    $ android --verbose create avd --name android_4_0_3 \
    --target android-15 --sdcard 32M
    ```

The following is a screenshot of the command output:

```
roman10@roman10-laptop:~$ android --verbose create avd --name android_4_0_3
--target android-15 --sdcard 32M
Auto-selecting single ABI armeabi-v7a
Android 4.0.3 is a basic Android platform.
Do you wish to create a custom hardware profile [no]
Created AVD 'android_4_0_3' based on Android 4.0.3, ARM (armeabi-v7a) proces
sor,
with the following hardware config:
hw.lcd.density=240
vm.heapSize=48
hw.ramSize=512
roman10@roman10-laptop:~$ █
```

10. Start the emulator, using the following command:

    ```
    $ emulator -wipe-data -avd android_4_0_3
    ```

    Alternatively, we can start the **Android Virtual Device Manager** window by using the command "`android avd`", and then choosing an emulator to launch, as follows:

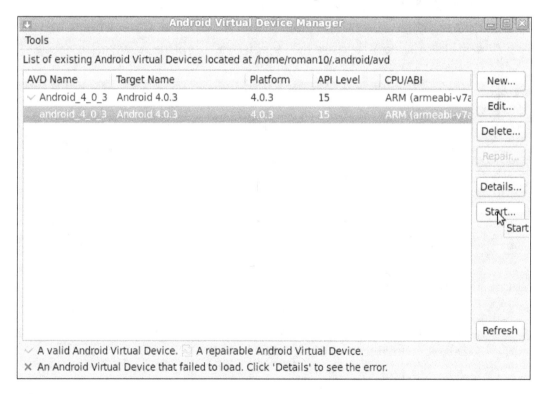

11. Install the app on the emulator. We first check the device serial number by using the following command:

    ```
    $ adb devices
    ```

    The following is a screenshot of the command output:

    ```
    roman10@roman10-laptop:~/Desktop/book-code/chapter3/HelloNDK$ adb devices
    List of devices attached
    emulator-5554   device
    ```

12. We then install the `debug.apk` file to the emulator by using the following command:

    ```
    $ adb -s emulator-5554 install bin/HelloNDK-debug.apk
    ```

    ```
    roman10@roman10-laptop:~/Desktop/book-code/chapter3/HelloNDK$ adb -s emulator-5554 install bin/HelloNDK-debug.apk
    465 KB/s (19611 bytes in 0.041s)
        pkg: /data/local/tmp/HelloNDK-debug.apk
    Success
    ```

 If only a single device is connected to the computer, there is no need to specify the device serial number. In the preceding commands , we can remove "-s emulator-5554".

13. Start the `HelloNDK` app on the emulator using the command in the following format:

    ```
    $ adb shell am start -n com.package.name/com.package.name.
    ActivityName
    ```

    In our example, we use the following ommand:

    ```
    $ adb -s emulator-5554 shell am start -n cookbook.chapter3/
    cookbook.chapter3.HelloNDKActivity
    ```

14. Run the app on a device.

    Suppose the device serial number is `HT21HTD09025`, then we can use the following command to install the app on an Android device.

    ```
    $ adb -s HT21HTD09025 install bin/HelloNDK-debug.apk
    ```

    In our example, we use the following command to start the app:

    ```
    $ adb -s HT21HTD09025 shell am start -n cookbook.chapter3/
    cookbook.chapter3.HelloNDKActivity
    ```

15. Create a release package.

Once we confirm that our application can run successfully, we may want to create a release package for uploading to the Android market. You can perform the following steps to achieve this:

1. Create a keystore. An Android app must be signed using a key from a keystore. A **keystore** is a collection of private keys. We can use the following command to create a keystore with a private key:

   ```
   $ keytool -genkey -v -keystore release_key.keystore \
   -alias androidkey \
   -keyalg RSA -keysize 2048 -validity 10000 \
   -dname "CN=MyCompany, OU=MyAndroidDev, O=MyOrg, L=Singapore,
   S=Singapore, C=65" \
   -storepass testkspw -keypass testkpw
   ```

   The following is a screenshot of the command output:

   ```
   roman10@roman10-laptop:~/Desktop/book-code/chapter3/HelloNDK$ keytool -genkey -v -keystore release_key.
   keystore -alias androidkey -keyalg RSA -keysize 2048 -validity 10000 -dname "CN=MyCompany, OU=MyAndroid
   Dev, O=MyOrg, L=Singapore, S=Singapore, C=65" -storepass testkspw -keypass testkpw
   Generating 2,048 bit RSA key pair and self-signed certificate (SHA1withRSA) with a validity of 10,000 d
   ays
           for: CN=MyCompany, OU=MyAndroidDev, O=MyOrg, L=Singapore, ST=Singapore, C=65
   [Storing release_key.keystore]
   roman10@roman10-laptop:~/Desktop/book-code/chapter3/HelloNDK$
   ```

   As shown, a keystore with password as `testkwpw` is created, and a RSA key pair with password as `testkpw` is added to the keystore.

2. Type the command "`ant release`" to build an `apk` for the app. The output can be found in the `bin` folder as `HelloNDK-release-unsigned.apk`.

3. Sign the `apk` by using the following command:

```
$ jarsigner -verbose -keystore <keystore name> -storepass <store
password> -keypass <key password> -signedjar <name of the signed
output> <unsigned input file name> <alias>
```

For our sample application, the command and output are as follows:

```
roman10@roman10-laptop:~/Desktop/book-code/chapter3/HelloNDK$ jarsigner -verbose -keystore
 release_key.keystore -storepass testkspw -keypass testkpw -signedjar bin/HelloNDK-release
-signed.apk bin/HelloNDK-release-unsigned.apk androidkey
   adding: META-INF/MANIFEST.MF
   adding: META-INF/ANDROIDK.SF
   adding: META-INF/ANDROIDK.RSA
  signing: res/layout/main.xml
  signing: AndroidManifest.xml
  signing: resources.arsc
  signing: res/drawable-hdpi/ic_launcher.png
  signing: res/drawable-ldpi/ic_launcher.png
  signing: res/drawable-mdpi/ic_launcher.png
  signing: classes.dex
  signing: lib/armeabi/libhello.so
```

4. Zip-align the `apk` file. The `zipalign` tool aligns the data inside an `apk` file
   for performance optimization. The following command can be used to align
   a signed `apk`:

```
$ zipalign -v 4 <app apk file name>   <aligned apk file name>
```

For our sample application, the command and output are as follows:

```
roman10@roman10-laptop:~/Desktop/book-code/chapter3/HelloNDK$ zipalign -v 4 bin/HelloNDK-r
elease-signed.apk bin/HelloNDK-release-aligned.apk
Verifying alignment of bin/HelloNDK-release-aligned.apk (4)...
      50 META-INF/MANIFEST.MF (OK - compressed)
     515 META-INF/ANDROIDK.SF (OK - compressed)
    1040 META-INF/ANDROIDK.RSA (OK - compressed)
    2166 res/layout/main.xml (OK - compressed)
    2545 AndroidManifest.xml (OK - compressed)
    3116 resources.arsc (OK)
    4476 res/drawable-hdpi/ic_launcher.png (OK)
    8508 res/drawable-ldpi/ic_launcher.png (OK)
   10108 res/drawable-mdpi/ic_launcher.png (OK)
   12349 classes.dex (OK - compressed)
   14023 lib/armeabi/libhello.so (OK - compressed)
Verification succesful
roman10@roman10-laptop:~/Desktop/book-code/chapter3/HelloNDK$ ▌
```

## How it works...

This recipe discusses how to build an Android NDK application from the command line.

Android NDK provides a build system with the following goals:

- **Simplicity**: It handles most of the heavy lifting stuff for developers, and we only need to write brief build files (`Android.mk` and `Application.mk`) to describe the sources need to be compiled.

- **Compatibility**: More build tools, platforms, and so on, may be added to NDK in future releases, but no changes are required for the build files.

Android NDK comes with a set of cross toolchains, including cross-compilers, cross-linkers, cross-assemblers, and so on. These tools can be found under `toolchains` folder of the NDK `root` directory. They can be used to generate binaries on different Android platforms (ARM, x86, or MIPS) on Linux, Mac OS, or Windows. Although it is possible to use the toolchains directly to build native code for Android, it is not recommended unless we're porting a project with its own build scripts. In this case, we may only need to change the original compiler to the NDK cross compiler to build it for Android.

In most cases, we'll describe the sources in `Android.mk` and specify the ABIs on `Application.mk`. Android NDK's `ndk-build` script will internally invoke the cross toolchain to build the native code for us. The following is a list of commonly used `ndk-build` options:

- `ndk-build`: It is used to build binaries.
- `ndk-build clean`: It cleans the generated binaries.
- `ndk-build V=1`: This builds binaries and displays the build commands. It is handy when we want to find out how things are built or checked for build bugs.
- `ndk-build -B`: This command forces a rebuild.
- `ndk-build NDK_DEBUG=1`: It generates debuggable build.
- `ndk-build NDK_DEBUG=0`: It generates a release build.

## There's more...

This recipe uses a lot of command-line tools in the Android SDK. This allows us to present complete instructions of how to create, build, and deploy an Android NDK project. However, we won't provide the details about these tools in this book since this book is dedicated to Android NDK. You may read more about those tools at `http://developer.android.com/tools/help/index.html`.

### Taking screenshots from the command line

Taking a screenshot from command line can be handy to record the display results for an automated test. However, Android does not provide a command-line tool to take a screenshot currently.

A Java program found at `\development\tools\screenshot\src\com\android\screenshot\` of the Android source code can be used to take screenshot. The code uses a similar method as the Eclipse DDMS plugin to take a screenshot, but from the command line. We incorporated the preceding code into an Eclipse Java project named `screenshot`, which can be downloaded from the website.

One can import the project and export an executable JAR file to use the tool. Suppose the exported JAR file is named `screenshot.jar`, then the following sample command uses it to take a screenshot from an emulator:

```
roman10@roman10-laptop:~/Desktop$ java -jar screenshot.jar -s emulator-5554 1.png
Taking screenshot from: emulator-5554
Success.
```

# Building an Android NDK application in Eclipse

The previous recipe discusses how to build an Android NDK application in the command line. This recipe demonstrates how to do it in the Eclipse IDE.

## Getting ready

Add NDK Preferences. Start Eclipse, then click on **Window | Preferences**. In the **Preferences** window, select **NDK** under **Android**. Click on **Browse** and select the NDK `root` folder. Click on **OK**.

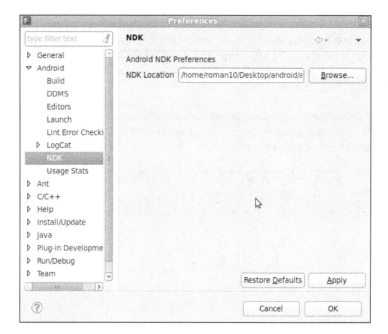

## How to do it...

The following steps create an NDK project using Eclipse:

1.  Create an Android application named `HelloNDKEclipse`. Set the package name as `cookbook.chapter3`. Create an activity named `HelloNDKEclipseActivity`. Please refer to the *Loading native libraries and registering native methods* recipe of *Chapter 2, Java Native Interface*, if you want more detailed instructions.

2.  Right-click on the project `HelloNDKEclipse`, select **Android Tools | Add Native Support**. A window similar to the following screenshot will appear. Click on **Finish** to dismiss it:

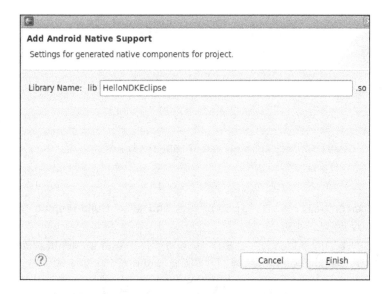

This will add a `jni` folder with two files (`HelloNDKEclipse.cpp` and `Android.mk`) inside, and switch Eclipse to C/C++ perspective.

3. Add the following content to `HelloNDKEclipse.cpp`:

```
#include <jni.h>

jstring getString(JNIEnv* env) {
   return env->NewStringUTF("Hello NDK");
}

extern "C" {
   JNIEXPORT jstring JNICALL Java_cookbook_chapter3_
HelloNDKEclipseActivity_getString(JNIEnv* env, jobject o){
      return getString(env);
   }
}
```

4. Change the content of HelloNDKEclipseActivity.java to below.

```
package cookbook.chapter3;

import android.os.Bundle;
import android.app.Activity;
import android.widget.TextView;

public class HelloNDKEclipseActivity extends Activity {
   @Override
   public void onCreate(Bundle savedInstanceState) {
```

```
        super.onCreate(savedInstanceState);
        TextView tv = new TextView(this);
        tv.setTextSize(30);
        tv.setText(getString());
        this.setContentView(tv);
    }
    public native String getString();
    static {
        System.loadLibrary("HelloNDKEclipse");
    }
}
```

5.  Right-click on `HelloNDKEclipse` project, and select **Build Project**. This will build the native library for us.

6.  Right-click on the project, go to **Run As**, and select **Android Application**. The phone screen will display something similar to the following screenshot:

## How it works...

This recipe discusses building the Android NDK application at Eclipse.

We have been using C in all previous recipes. Starting from this recipe, we'll be writing our code in C++.

By default, Android provides minimal C++ support. There's no **Run-time Type Information** (**RTTI**) and C++ exceptions support, and even the C++ standard library support, is partial. The following is a list of the C++ headers supported by Android NDK by default:

```
cassert, cctype, cerrno, cfloat, climits, cmath, csetjmp, csignal,
cstddef, cstdint, cstdio, cstdlib, cstring, ctime, cwchar, new, stl_
pair.h, typeinfo, utility
```

It is possible to add more C++ support by using different C++ libraries. NDK comes with the `gabi++`, `stlport`, and `gnustl` C++ libraries, besides the system default one.

In our sample code, we used an external "C" to wrap the C++ method. This is to avoid C++ mangling of the JNI function names. C++ name mangling could change the function names to include type information about parameters, whether the function is virtual or not, and so on. While this enables C++ to link overloaded functions, it breaks the JNI function discovery mechanism.

We can also use the explicit function registration method covered in the *Loading native libraries and registering native methods* recipe of *Chapter 2, Java Native Interface*, to get rid of the wrapping.

# Building an Android NDK application for different ABIs

Native code is compiled into binaries. Therefore, one set of binaries can only run on a specific architecture. Android NDK comes with techniques and tools to allow developers to compile the same source code for multiple architectures easily.

## Getting ready

An **Application Binary Interface** (**ABI**) defines how the Android application's machine code is supposed to interact with the system at runtime, including the CPU instruction set, endianness, alignment of memory, and so on. An ABI basically defines a type of architecture.

The following table briefly summarizes the four ABIs supported by Android:

| ABI name | Support | Not support | Optional |
|---|---|---|---|
| armeabi | ▸ ARMv5TE instruction set<br>▸ Thumb (also known as Thumb-1) instructions | Hardware-assisted floating point computation | |
| armeabi-v7a | ▸ Whatever is supported in `armeabi`<br>▸ VFP hardware FPU instructions<br>▸ Thumb-2 instruction set<br>▸ VFPv3-D16 is used. | | ▸ Advanced SIMD (also known as NEON)<br>▸ VFPv3-D32<br>▸ ThumbEE |
| x86 | ▸ Instruction set commonly known as "x86" or "IA-32".<br>▸ MMX, SSE, SSE2, and SSE3 instruction set extensions | | ▸ MOVBE instruction<br>▸ SSSE3 "supplemental SSE3" extension<br>▸ Any variant of "SSE4" |
| mips | ▸ MIPS32r1 instruction set<br>▸ Hard-Float<br>▸ O32 | ▸ DSP application specific extension<br>▸ MIPS16<br>▸ micromips | |

armeabi and armeabi-v7a are the two most commonly used ABIs for Android devices. ABI armeabi-v7a is compatible with armeabi, which means applications compiled for armeabi can run on armeabi-v7a too. But the reverse is not true, since armeabi-v7a includes additional features. In the following section, we briefly introduce some technical terms referred to frequently in armeabi and armeabi-v7a:

> **Thumb**: This instruction set consists of 16-bit instructions, which is a subset of the 32-bit instruction set of the standard ARM. Some instructions in the 32-bit instruction set are not available for Thumb, but can be simulated with several Thumb instructions. The narrower 16-bit instruction set can offer memory advantages.
>
> Thumb-2 extends Thumb-1 by adding some 32-bit instructions, which results in a variable-length instruction set. Thumb-2 aims to attain code density like to Thumb-1 and performance similar to standard ARM instruction set on a 32-bit memory.
>
> Android NDK generates the thumb code by default, unless LOCAL_ARM_MODE is defined in the Android.mk file.

> **Vector Floating Point (VFP)**: It is an extension to the ARM processor, which provides low cost floating point computation.

> **VFPv3-D16 and VFPv3-D32**: VFPv3-D16 refers to 16 dedicated 64-bit floating point registers. Similarly, VFPv3-D32 means there are 32 64-bit floating point registers. These registers speed up floating point computation.

> **NEON**: NEON is the nickname for the ARM **Advanced Single Instruction Multiple Data** (**SIMD**) instruction set extension. It requires VFPv3-D32, which means 32 hardware FPU 64-bit registers will be used. It provides a set of scalar/vector instructions and registers, which are comparable to MMX/SSE/SDNow! in the x86 world. It is not supported by all Android devices, but many new devices have NEON support. NEON can accelerate media and signal processing applications significantly by executing up to 16 operations simultaneously.

One can refer to ARM documentation website at http://infocenter.arm.com/help/index.jsp for more detailed information. We don't discuss x86 and mips ABI here, because few Android devices run on these architecture.

Read the *Building Android NDK Application at Eclipse* recipe before going through this one.

## How to do it...

The following steps build an Android project for different ABIs:

1.  Create an Android application named HelloNDKMultipleABI. Set the package name as cookbook.chapter3. Create an activity named HelloNDKMultipleABIActivity.

2. Right-click on the `HelloNDKMultipleABI` project, select **Android Tools | Add Native Support**. A window appears, click on **Finish** to dismiss it. This will add a `jni` folder with two files (`HelloNDKMultipleABI.cpp` and `Android.mk`) inside, and switch Eclipse to the C/C++ perspective.

3. Add the following content to the `HelloNDKMultipleABI.cpp` file:

```
#include <jni.h>

jstring getString(JNIEnv* env) {
   return env->NewStringUTF("Hello NDK");
}

extern "C" {
  JNIEXPORT jstring JNICALL Java_cookbook_chapter3_
HelloNDKMultipleABIActivity_getString(JNIEnv* env, jobject o){
    return getString(env);
  }
}
```

4. Change the `HelloNDKMultipleABIActivity.java` file to the following content:

```
package cookbook.chapter3;

import android.os.Bundle;
import android.app.Activity;
import android.widget.TextView;

public class HelloNDKMultipleABIActivity extends Activity {

    @Override
    public void onCreate(Bundle savedInstanceState) {
        super.onCreate(savedInstanceState);
        TextView tv = new TextView(this);
        tv.setTextSize(30);
        tv.setText(getString());
        this.setContentView(tv);
    }
    public native String getString();
    static {
        System.loadLibrary("HelloNDKMultipleABI");
    }
}
```

5. Add a new file named `Application.mk` under the project's `jni` folder with the following content:

   ```
   APP_ABI := armeabi armeabi-v7a
   ```

6. Right-click on the `HelloNDKMultipleABIActivity` project, and select **Build Project**. This will build the native library for us.

7. Create two emulators, with ABI set to `armeabi` and `armeabi-v7a` respectively. The following screenshot depicts how an emulator is created with the `armeabi` ABI:

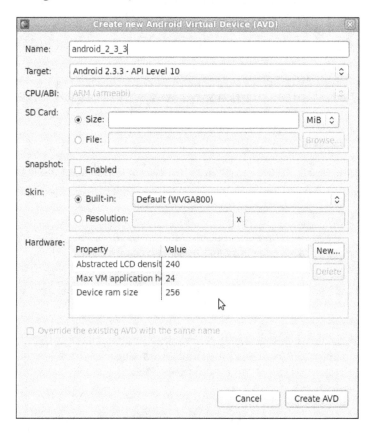

8. Run the sample Android application on the two emulators. The same result is shown on both of them:

9.  Change the content of `Application.mk` to the following code snippet and run the sample application on the two emulators. The application will still work on both the emulators:

    ```
    #APP_ABI := armeabi armeabi-v7a
    APP_ABI := armeabi
    ```

10. Change the content of `Application.mk` as follows:

    ```
    #APP_ABI := armeabi armeabi-v7a
    #APP_ABI := armeabi
    APP_ABI := armeabi-v7a
    ```

11. Run the sample application on the two emulators. The application works on the `armeabi-v7a` emulator, but crashes on `armeabi` emulator, as shown in the following screenshot:

## How it works...

An Android device can define one or two ABIs. For typical x86-, MIPS-, ARMv5-, and ARMv6-based devices, there's only a primary ABI. Based on the platform, it can be x86, mips, or armeabi. For a typical ARMv7-based device, the primary ABI is usually armeabi-v7a, and it also has a secondary ABI as armeabi. This enables binaries compiled for either armeabi or armeabi-v7a to run on ARMv7 devices. In our example, we demonstrated that the app can work on both armeabi and armeabi-v7a emulators when built against only armeabi.

At installation, the Android package manager searches for native libraries built for the primary ABI and copies them to the application's data directory. If not found, it then searches the native libraries built for the secondary ABI. This ensures that only the proper native libraries are installed.

In our example, when we compile the binary against armeabi-v7a only, the native library won't get installed on the armeabi emulator, subsequently the native library cannot be loaded, and a crash will be shown.

# Building Android NDK applications for different CPU features

Many projects use native code to improve performance. One advantage of developing in NDK over SDK is that we can build different packages for different CPUs, which is the topic of this recipe.

## Getting ready

Please read the *Building Android NDK application for different ABIs* recipe before going through this one.

## How to do it...

The following steps build Android NDK applications for different CPU features.

1. At Eclipse, click on **File | New | Other**. Select **Android Project** from **Existing Code** under **Android** as shown in the following screenshot. Then click on **Next**:

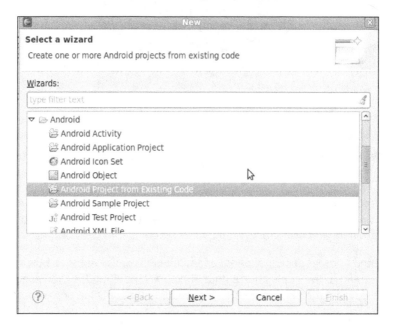

2. Browse to the `samples/hello-neon` folder of the Android NDK folder. Then click on **Finish**.

3. Start a terminal, then go to the `samples/hello-neon/jni` folder. Type the command "`ndk-build`" to build the binaries.

4. Run the Android project on different devices and emulators. Based on your device/ emulator ABI and availability of the NEON feature, you should be able to see the results as follows:

   ❑ For Android device with armeabi ABI, the result is as follows:

```
HelloNeon
FIR Filter benchmark:
C version      : 2238.97 ms
Neon version   : Not an ARMv7 CPU !
```

   ❑ For Android device with armeabi-v7a ABI and NEON, the result is as follows:

```
HelloNeon
FIR Filter benchmark:
C version      : 365.631 ms
Neon version   : 178.589 ms (x2.04733 faster)
```

## How it works...

Android devices are roughly divided by ABIs. However, different devices with the same ABI can have different CPU extensions and features. These extensions and features are optional and therefore we don't know whether a user's device has them until runtime. Detecting and making use of these features can sometimes improve app performance significantly on certain devices.

Android NDK contains a library named `cpufeatures`, which can be used to detect the CPU family and optional features at runtime. As illustrated in the sample code, the following steps indicate how to use this library:

1. Add it in the static library list in `Android.mk` as follows:

   `LOCAL_STATIC_LIBRARIES := cpufeatures`

2. At the end of the `Android.mk` file, import the `cpufeatures` module:

   `$(call import-module,cpufeatures)`

3. In the code, include the header file `<cpu-features.h>`.

4. Call detection functions; Currently `cpufeatures` provides only three functions:

5. Get the CPU family. The function prototype is as follows:

   `AndroidCpuFamily    android_getCpuFamily();`

It returns an enum. The supported CPU families are listed in the section to follow.

```
ANDROID_CPU_FAMILY_MIPS
ANDROID_CPU_FAMILY_MIPS
ANDROID_CPU_FAMILY_ARM
```

6. Get the optional CPU features. Each CPU feature is represented by a bit flag and the bit is set to 1 if the feature is available. The function prototype is as follows:

```
uint64_t    android_getCpuFeatures();
```

For the ARM CPU family, the supported CPU feature detections are as follows:

- ANDROID_CPU_ARM_FEATURE_ARMv7: It means that the ARMv7-a instruction is supported.

- ANDROID_CPU_ARM_FEATURE_VFPv3: It means that the VFPv3 hardware FPU instruction set extension is supported. Note that this refers to VFPv3-D16, which provides 16 hardware FP registers.

- ANDROID_CPU_ARM_FEATURE_NEON: It means that he ARM Advanced SIMD (also known as NEON) vector instruction set extension is supported. Note that such CPUs also support VFPv3-D32, which provides 32 hardware FP registers.

For the x86 CPU family, the supported CPU feature detections are as follows:

- ANDROID_CPU_X86_FEATURE_SSSE3: It means that the SSSE3 instruction extension set is supported.

- ANDROID_CPU_X86_FEATURE_POPCNT: It means that the POPCNT instruction is supported.

- ANDROID_CPU_X86_FEATURE_MOVBE: It means that the MOVBE instruction is supported.

We can do a "&" operation to detect if a feature is available or not, as follows:

```
uint64_t features = android_getCpuFeatures();
if ((features & ANDROID_CPU_ARM_FEATURE_NEON) == 0) {
  //NEON is not available
} else {
  //NEON is available
}
```

Get the number of CPU cores on the device:

```
int            android_getCpuCount(void);
```

Since NDK r8c, more CPU feature detections are available. Please refer to `sources/android/cpufeatures/cpu-features.c` for more details.

## There's more...

There are a few more noteworthy points about CPU features on Android.

### More about CPU feature detection

The `cpufeatures` library can only detect a limited set of CPU features. It is possible to implement our own CPU detection mechanisms. By looking at the NDK source code at `/sources/android/cpufeatures/`, one can find that the `cpufeatures` library essentially looks at the `/proc/cpuinfo` file. We can read this file and parse the content in our application. The following is a screenshot of the file content:

```
roman10@roman10-laptop:~/Desktop/ndk/sources/android/cpufeatures$ adb -d shell
# cat /proc/cpuinfo
Processor       : ARMv7 Processor rev 1 (v7l)
BogoMIPS        : 163.57
Features        : swp half thumb fastmult vfp edsp thumbee neon vfpv3
CPU implementer : 0x51
CPU architecture: 7
CPU variant     : 0x1
CPU part        : 0x00f
CPU revision    : 1

Hardware        : vivo
Revision        : 0080
Serial          : 0000000000000000
# exit;
```

Please refer to the Android project `cpuinfo`, available in the book's website for how to do this programmatically.

### Different approaches of building for different CPU features

There are a few approaches to building native code for different CPU features:

▶ **Single library, different binaries at build time**: This is demonstrated in the sample project. The `helloneon-intrinsics.c` file is only compiled for armeabi-v7a ABI.

▶ **Single library, different execution paths at runtime**: This is also shown in the sample project. The code detects whether the NEON feature is available or not at runtime and executes different code blocks.

▶ **Different libraries, load appropriate library at runtime**: Sometimes, we may want to compile the source code into different libraries and differentiate them by names. For example, we may have `libmylib-neon.so` and `libmylib-vfpv3.so`. We detect the CPU feature at runtime and load the appropriate library.

▶ **Different packages, load appropriate library at runtime**: If the library is big, it is desirable to deploy different binaries for different CPUs as separate packages. This is done by many video players available on Google Play (for example, MX Player).

# Debugging an Android NDK application with logging messages

Android logging system provides a method for collecting logs from various applications into a series of circular buffers. The `logcat` command is used to view the logs. Log message is the simplest method of debugging a program, yet one of the most powerful ones. This recipe focuses on message logging in NDK.

## How to do it...

The following steps create our sample Android project:

1. Create an Android application named `NDKLoggingDemo`. Set the package name as `cookbook.chapter3`. Create an activity named `NDKLoggingDemoActivity`. Please refer to the *Loading native libraries and registering native methods* recipe of *Chapter 2*, *Java Native Interface*, if you want more detailed instructions.

2. Right-click on the project `NDKLoggingDemo`, select **Android Tools | Add Native Support**. A window appears, click on **Finish** to dismiss it.

3. Add a new file named `mylog.h` under the `jni` folder, and add the following content to it:

```
#ifndef COOKBOOK_LOG_H
#define COOKBOOK_LOG_H

#include <android/log.h>

#define LOG_LEVEL 9
#define LOG_TAG "NDKLoggingDemo"

#define LOGU(level, ...) if (level <= LOG_LEVEL) {__android_log_
print(ANDROID_LOG_UNKNOWN, LOG_TAG, __VA_ARGS__);}
#define LOGD(level, ...) if (level <= LOG_LEVEL) {__android_log_
print(ANDROID_LOG_DEFAULT, LOG_TAG, __VA_ARGS__);}
#define LOGV(level, ...) if (level <= LOG_LEVEL) {__android_log_
print(ANDROID_LOG_VERBOSE, LOG_TAG, __VA_ARGS__);}
```

```
#define LOGDE(level, ...) if (level <= LOG_LEVEL) {__android_log_
print(ANDROID_LOG_DEBUG, LOG_TAG, __VA_ARGS__);}
#define LOGI(level, ...) if (level <= LOG_LEVEL) {__android_log_
print(ANDROID_LOG_INFO, LOG_TAG, __VA_ARGS__);}
#define LOGW(level, ...) if (level <= LOG_LEVEL) {__android_log_
print(ANDROID_LOG_WARN, LOG_TAG, __VA_ARGS__);}
#define LOGE(level, ...) if (level <= LOG_LEVEL) {__android_log_
print(ANDROID_LOG_ERROR, LOG_TAG, __VA_ARGS__);}
#define LOGF(level, ...) if (level <= LOG_LEVEL) {__android_log_
print(ANDROID_LOG_FATAL, LOG_TAG, __VA_ARGS__);}
#define LOGS(level, ...) if (level <= LOG_LEVEL) {__android_log_
print(ANDROID_LOG_SILENT, LOG_TAG, __VA_ARGS__);}

#endif
```

4.  Add the following content to NDKLoggingDemo.cpp:

```cpp
#include <jni.h>
#include "mylog.h"

void outputLogs() {
  LOGU(9, "unknown log message");
  LOGD(8, "default log message");
  LOGV(7, "verbose log message");
  LOGDE(6, "debug log message");
  LOGI(5, "information log message");
  LOGW(4, "warning log message");
  LOGE(3, "error log message");
  LOGF(2, "fatal error log message");
  LOGS(1, "silent log message");
}

extern "C" {
  JNIEXPORT void JNICALL Java_cookbook_chapter3_
NDKLoggingDemoActivity_LoggingDemo(JNIEnv* env, jobject o){
    outputLogs();
  }
}
```

5.  Change the content of NDKLoggingDemoActivity.java to the following:

```java
package cookbook.chapter3;

import android.os.Bundle;
import android.app.Activity;

public class NDKLoggingDemoActivity extends Activity {
```

```
        @Override
        public void onCreate(Bundle savedInstanceState) {
            super.onCreate(savedInstanceState);
            LoggingDemo();
        }
        public native void LoggingDemo();
        static {
            System.loadLibrary("NDKLoggingDemo");
        }
    }
```

6. Change the `Android.mk` file to include the Android log library as follows:

```
LOCAL_PATH := $(call my-dir)

include $(CLEAR_VARS)

LOCAL_MODULE    := NDKLoggingDemo
LOCAL_SRC_FILES := NDKLoggingDemo.cpp
LOCAL_LDLIBS := -llog
include $(BUILD_SHARED_LIBRARY)
```

7. Right-click on the `NDKLoggingDemo` project, and select **Build Project**.

8. Start the monitor logcat output by entering the following command. Then, start the sample Android app on an Android device:

   **$ adb logcat -c**

   **$ adb logcat NDKLoggingDemo:I *:S -v time**

   The following is a screenshot of the logcat output:

```
roman10@roman10-laptop:~/Desktop$ adb logcat NDKLoggingDemo:I *:S -v time
--------- beginning of /dev/log/main
--------- beginning of /dev/log/system
I/NDKLoggingDemo(25969): information log message
W/NDKLoggingDemo(25969): warning log message
E/NDKLoggingDemo(25969): error log message
F/NDKLoggingDemo(25969): fatal error log message
S/NDKLoggingDemo(25969): silent log message
```

9. Start another command line terminal, and enter the following command in it:

   **$ adb logcat NDKLoggingDemo:V *:S -v time**

This will result into the following output:

```
roman10@roman10-laptop:~/Desktop$ adb logcat NDKLoggingDemo:V *:S -v time
--------- beginning of /dev/log/main
--------- beginning of /dev/log/system
V/NDKLoggingDemo(25969): verbose log message
D/NDKLoggingDemo(25969): debug log message
I/NDKLoggingDemo(25969): information log message
W/NDKLoggingDemo(25969): warning log message
E/NDKLoggingDemo(25969): error log message
F/NDKLoggingDemo(25969): fatal error log message
S/NDKLoggingDemo(25969): silent log message
```

10. Change the line in `mylog.h` from `#define LOG_LEVEL 9` to `#define LOG_LEVEL 4`. Rebuild the application, then restart the application.

11. The outputs of the two terminals we started earlier are the same.

```
W/NDKLoggingDemo(26151): warning log message
E/NDKLoggingDemo(26151): error log message
F/NDKLoggingDemo(26151): fatal error log message
S/NDKLoggingDemo(26151): silent log message
```

## How it works...

This recipe shows how to use Android log messages. Each log message in Android consists of the following three parts:

- **Priority**: It is usually used to filter log messages. In our project, we can control the log by changing the following code:

  `#define LOG_LEVEL 4`

  Alternatively, we can selectively display the log output using `logcat`.

- **Log tag**: It is usually used to identify the log source.

- **Log message**: It provides the detailed log message.

Sending log messages on Android consumes CPU resources and frequent log messages can affect the application performance. In addition, the logs are stored in a circular buffer. Too many logs will overwrite some earlier logs, which may not be desirable. Due to these facts, it is recommended we only log errors and exceptions at the release build.

`logcat` is the command-line tool to view Android logs. It allows one to filter logs according to the log tag and priority. It is also capable of dispalying logs in different formats.

For example, we used the following `logcat` command in step 8 of the preceding *How to do it...* section.

```
adb logcat NDKLoggingDemo:I *:S -v time
```

The command filters out logs except those with the `NDKLoggingDemo` tag and priority `I` (information) or higher. The filter is given in a `tag:priority` format. `NDKLoggingDemo:I` indicates logs with a `NDKLoggingDemo` tag and priority information or higher will be displayed. `*:S` sets the priority level for all other tags as "silent".

More details about logcat filtering and format can be found at `http://developer.android.com/tools/help/logcat.html` and `http://developer.android.com/tools/debugging/debugging-log.html#outputFormat`.

# Debugging an Android NDK application with CheckJNI

JNI does little error checking for better performance. As a result, errors usually lead to a crash. A mode called `CheckJNI` is offered by Android. In this mode, a set of JNI functions with extended checks are called instead of the normal JNI functions. This recipe discusses how to enable the CheckJNI mode to debug Android NDK applications.

## How to do it...

The following steps create a sample Android project and enable the `CheckJNI` mode:

1. Create an Android application named `CheckJNIDemo`. Set the package name as `cookbook.chapter3`. Create an activity named `CheckJNIDemoActivity`. Please refer to the *Loading native libraries and registering native methods* recipe of *Chapter 2, Java Native Interface*, if you want more detailed instructions.

2. Right-click on the project `CheckJNIDemo`, select **Android Tools | Add Native Support**. A window appears; click on **Finish** to dismiss it.

3. Add the following content to `CheckJNIDemo.cpp`.

4. Change `CheckJNIDemoActivity.java` to the following:

```
package cookbook.chapter3;
import android.os.Bundle;
import android.app.Activity;

public class CheckJNIDemoActivity extends Activity {
    @Override
    public void onCreate(Bundle savedInstanceState) {
        super.onCreate(savedInstanceState);
        setContentView(R.layout.activity_check_jnidemo);
```

```
        CheckJNIDemo();
    }
    public native int[] CheckJNIDemo();
    static {
        System.loadLibrary("CheckJNIDemo");
    }
}
```

5.  Right-click on `CheckJNIDemo` project, and select **Build Project**.

6.  Start the monitor logcat output by entering "`adb logcat -v time`" on a command-line console. Then, start the sample Android app on an Android device. The application will crash, and the logcat output will be displayed as follows:

```
08-19 14:54:13.809 D/AndroidRuntime(23772): Shutting down VM
08-19 14:54:13.809 W/dalvikvm(23772): threadid=1: thread exiting with uncaught exception (group=0x40015560)
08-19 14:54:13.809 E/AndroidRuntime(23772): FATAL EXCEPTION: main
08-19 14:54:13.809 E/AndroidRuntime(23772): java.lang.OutOfMemoryError: array size too large
08-19 14:54:13.809 E/AndroidRuntime(23772):     at cookbook.chapter3.CheckJNIDemoActivity.CheckJNIDemo(Native Met
hod)
08-19 14:54:13.809 E/AndroidRuntime(23772):     at cookbook.chapter3.CheckJNIDemoActivity.onCreate(CheckJNIDemoAc
tivity.java:12)
08-19 14:54:13.809 E/AndroidRuntime(23772):     at android.app.Instrumentation.callActivityOnCreate(Instrumentati
on.java:1047)
08-19 14:54:13.809 E/AndroidRuntime(23772):     at android.app.ActivityThread.performLaunchActivity(ActivityThrea
d.java:1722)
08-19 14:54:13.809 E/AndroidRuntime(23772):     at android.app.ActivityThread.handleLaunchActivity(ActivityThread
.java:1784)
08-19 14:54:13.809 E/AndroidRuntime(23772):     at android.app.ActivityThread.access$1500(ActivityThread.java:123
)
08-19 14:54:13.809 E/AndroidRuntime(23772):     at android.app.ActivityThread$H.handleMessage(ActivityThread.java
:939)
08-19 14:54:13.809 E/AndroidRuntime(23772):     at android.os.Handler.dispatchMessage(Handler.java:99)
08-19 14:54:13.809 E/AndroidRuntime(23772):     at android.os.Looper.loop(Looper.java:130)
08-19 14:54:13.809 E/AndroidRuntime(23772):     at android.app.ActivityThread.main(ActivityThread.java:3835)
08-19 14:54:13.809 E/AndroidRuntime(23772):     at java.lang.reflect.Method.invokeNative(Native Method)
08-19 14:54:13.809 E/AndroidRuntime(23772):     at java.lang.reflect.Method.invoke(Method.java:507)
08-19 14:54:13.809 E/AndroidRuntime(23772):     at com.android.internal.os.ZygoteInit$MethodAndArgsCaller.run(Zyg
oteInit.java:847)
08-19 14:54:13.809 E/AndroidRuntime(23772):     at com.android.internal.os.ZygoteInit.main(ZygoteInit.java:605)
08-19 14:54:13.809 E/AndroidRuntime(23772):     at dalvik.system.NativeStart.main(Native Method)
08-19 14:54:13.819 W/ActivityManager( 1429):  Force finishing activity cookbook.chapter3/.CheckJNIDemoActivity
```

7.  Enable CheckJNI.

    ❏   When the emulator is being used by you, the CheckJNI is on by default.

    ❏   If you're using a rooted device, the following sequence of commands can be used to restart the runtime with CheckJNI enabled. The commands stop the running Android instance, change the system properties to enable CheckJNI, and then restart Android.

        `$ adb shell stop`

        `$ adb shell setprop dalvik.vm.checkjni true`
        `$ adb shell start`

    ❏   If you have a regular device, you can use the following command:

        `$ adb shell setprop debug.checkjni 1`

8. Run the Android application again. The logcat output will be displayed as follows:

```
08-19 15:13:28.856 D/dalvikvm(23924): Late-enabling CheckJNI
08-19 15:13:28.916 D/szipinf (23924): Initializing inflate state
08-19 15:13:28.926 D/dalvikvm(23924): Trying to load lib /data/data/cookbook.chapter3/lib/libCheckJNIDemo.so 0x4
05137b0
08-19 15:13:28.926 D/dalvikvm(23924): Added shared lib /data/data/cookbook.chapter3/lib/libCheckJNIDemo.so 0x405
137b0
08-19 15:13:28.926 D/dalvikvm(23924): No JNI_OnLoad found in /data/data/cookbook.chapter3/lib/libCheckJNIDemo.so
 0x405137b0, skipping init
08-19 15:13:28.976 D/dalvikvm(23924): GC_EXTERNAL_ALLOC freed 46K, 51% free 2688K/5379K, external 0K/0K, paused
36ms
08-19 15:13:29.006 W/dalvikvm(23924): JNI WARNING: negative length for array allocation (Check_NewIntArray)
08-19 15:13:29.006 I/dalvikvm(23924): "main" prio=5 tid=1 NATIVE
08-19 15:13:29.006 I/dalvikvm(23924):   | group="main" sCount=0 dsCount=0 obj=0x4001f1a0 self=0xce68
08-19 15:13:29.006 I/dalvikvm(23924):   | sysTid=23924 nice=0 sched=0/0 cgrp=default handle=-1345002432
08-19 15:13:29.006 I/dalvikvm(23924):   | schedstat=( 76568606 66436763 100 )
08-19 15:13:29.006 I/dalvikvm(23924):   at cookbook.chapter3.CheckJNIDemoActivity.CheckJNIDemo(Native Method)
08-19 15:13:29.006 I/dalvikvm(23924):   at cookbook.chapter3.CheckJNIDemoActivity.onCreate(CheckJNIDemoActivity.
java:12)
08-19 15:13:29.006 I/dalvikvm(23924):   at android.app.Instrumentation.callActivityOnCreate(Instrumentation.java
:1047)
08-19 15:13:29.006 I/dalvikvm(23924):   at android.app.ActivityThread.performLaunchActivity(ActivityThread.java:
1722)
08-19 15:13:29.006 I/dalvikvm(23924):   at android.app.ActivityThread.handleLaunchActivity(ActivityThread.java:1
784)
08-19 15:13:29.006 I/dalvikvm(23924):   at android.app.ActivityThread.access$1500(ActivityThread.java:123)
08-19 15:13:29.006 I/dalvikvm(23924):   at android.app.ActivityThread$H.handleMessage(ActivityThread.java:939)
08-19 15:13:29.006 I/dalvikvm(23924):   at android.os.Handler.dispatchMessage(Handler.java:99)
08-19 15:13:29.006 I/dalvikvm(23924):   at android.os.Looper.loop(Looper.java:130)
08-19 15:13:29.006 I/dalvikvm(23924):   at android.app.ActivityThread.main(ActivityThread.java:3835)
08-19 15:13:29.006 I/dalvikvm(23924):   at java.lang.reflect.Method.invokeNative(Native Method)
08-19 15:13:29.006 I/dalvikvm(23924):   at java.lang.reflect.Method.invoke(Method.java:507)
08-19 15:13:29.006 I/dalvikvm(23924):   at com.android.internal.os.ZygoteInit$MethodAndArgsCaller.run(ZygoteInit
.java:847)
08-19 15:13:29.006 I/dalvikvm(23924):   at com.android.internal.os.ZygoteInit.main(ZygoteInit.java:605)
08-19 15:13:29.006 I/dalvikvm(23924):   at dalvik.system.NativeStart.main(Native Method)
08-19 15:13:29.006 I/dalvikvm(23924):
08-19 15:13:29.006 E/dalvikvm(23924): VM aborting
```

## How it works...

The CheckJNI mode uses a set of JNI functions, which have more error checking than the default one. This makes it easier to find JNI programming bugs. The CheckJNI mode currently checks the following errors:

▶ **Negative-sized array**: It attempts to allocate an array of negative size.

▶ **Bad reference**: It passes a bad reference `jarray/jclass/jobject/jstring` to a JNI function. Passing `NULL` to JNI function expecting a non-NULL argument.

▶ **Class names**: It passes the class names of invalid style to the JNI function. Valid class names are separate by "/" as in "`java/lang/String`".

▶ **Critical calls**: It calls a JNI function between a "critical" get function and its corresponding release.

▶ **Exceptions**: It calls a JNI function when there's a pending exception.

▶ **jfieldIDs**: It invalidates `jfieldIDs` or assigns `jfieldIDs` from one type to another.

▶ **jmethodIDs**: It's similar to jfieldIDs.

- ▶ **References**: It uses `DeleteGlobalRef/DeleteLocalRef` on references of wrong types.
- ▶ **Release mode**: It passes a release mode other than `0`, `JNI_ABORT`, and `JNI_COMMIT` to a release call.
- ▶ **Type safety**: It returns an incompatible type from a native method.
- ▶ **UTF-8**: It passes invalid modified UTF-8 string to JNI functions.

More error checking may be added to CheckJNI as Android evolves. Currently, the following checks are not supported:

- ▶ Misuse of local references

# Debugging an Android NDK application with NDK GDB

Android NDK introduces a shell script named `ndk-gdb` to help one to launch a debugging session to debug the native code.

## Getting ready

The project must meet the following requirements in order to debug it with `ndk-gdb`:

- ▶ The application is built with the `ndk-build` command.
- ▶ `AndroidManifest.xml` has the `android:debuggable` attribute of the `<application>` element set to `true`. This indicates that the application is debuggable even when it is running on a device in the user mode.
- ▶ The application should be running on Android 2.2 or higher.

Please read the *Building Android NDK Application at Eclipse* recipe before going through this one.

## How to do it...

The following steps create a sample Android project and debug it using NDK GDB.

1. Create an Android application named `HelloNDKGDB`. Set the package name as `cookbook.chapter3`. Create an activity named `HelloNDKGDBActivity`. Please refer to the *Loading native libraries and registering native methods* recipe of *Chapter 2*, *Java Native Interface*, if you want more detailed instructions.

2. Right-click on the project `HelloNDKGDB`, select **Android Tools | Add Native Support**. A window appears; click on **Finish** to dismiss it.

3. Add the following code to `HelloNDKGDB.cpp`:

```cpp
#include <jni.h>
#include <unistd.h>

int multiply(int i, int j) {
  int x = i * j;
  return x;
}


extern "C" {
  JNIEXPORT jint JNICALL Java_cookbook_chapter3_
HelloNDKGDBActivity_multiply(JNIEnv* env, jobject o, jint pi, jint
pj){
    int i = 1, j = 0;
    while (i) {
      j=(++j)/100;

    }
    return multiply(pi, pj);
  }
}
```

4. Change the content of `HelloNDKGDBActivity.java` to the following:

```java
package cookbook.chapter3;

import android.os.Bundle;
import android.widget.TextView;
import android.app.Activity;

public class HelloNDKGDBActivity extends Activity {

    @Override
    public void onCreate(Bundle savedInstanceState) {
        super.onCreate(savedInstanceState);
        TextView tv = new TextView(this);
        tv.setTextSize(30);
        tv.setText("10 x 20 = " + multiply(10, 20));
        this.setContentView(tv);
    }
    public native int multiply(int a, int b);
    static {
        System.loadLibrary("HelloNDKGDB");
    }
}
```

5. Make sure that the `debuggable` attribute in `AndroidManifest.xml` is set to `true`. The following code snippet is a part of the application element extracted from `AndroidManifest.xml` of our sample project:

```
<application
        android:icon="@drawable/ic_launcher"
        android:label="@string/app_name"
        android:theme="@style/AppTheme"
        android:debuggable="true"
        >
```

6. Build the native library with the command "`ndk-build NDK_DEBUG=1`". Alternatively, we can configure the `build` command at Eclipse under **C/C++ Build** of the project **Properties**. This is demonstrated in the *Debugging Android NDK application at Eclipse* recipe .

7. Run the application on an Android device. Then, start a terminal and enter the following command:

```
$ ndk-gdb
```

8. Once the debugger is attached to the remote process, we can issue GDB commands to start debugging the app. This is shown as follows:

```
roman10@roman10-laptop:~/Desktop/book-code/HelloNDKGDB$ ndk-gdb
GNU gdb (GDB) 7.3.1-gg2
Copyright (C) 2011 Free Software Foundation, Inc.
License GPLv3+: GNU GPL version 3 or later <http://gnu.org/licenses/gpl.html>
This is free software: you are free to change and redistribute it.
There is NO WARRANTY, to the extent permitted by law.  Type "show copying"
and "show warranty" for details.
This GDB was configured as "--host=x86_64-linux-gnu --target=arm-linux-android".
For bug reporting instructions, please see:
<http://www.gnu.org/software/gdb/bugs/>.
warning: while parsing target library list (at line 2): No segment defined for /
system/bin/linker
BFD: /home/roman10/Desktop/book-code/HelloNDKGDB/obj/local/armeabi/linker: warni
ng: sh_link not set for section `.ARM.exidx'
warning: Could not load shared library symbols for 54 libraries, e.g. libstdc++.
so.
Use the "info sharedlibrary" command to see the complete listing.
Do you need "set solib-search-path" or "set sysroot"?
0x80400dd2 in Java_cookbook_chapter3_HelloNDKGDBActivity_multiply (env=0xabe0, o
=0x4051b718, pi=10, pj=20) at jni/HelloNDKGDB.cpp:13
13                              j=(++j)/100;
(gdb) info break
No breakpoints or watchpoints.
(gdb) b 6
Breakpoint 1 at 0x80400da6: file jni/HelloNDKGDB.cpp, line 6.
(gdb) set var i=0
(gdb) c
Continuing.

Breakpoint 1, multiply (i=10, j=20) at jni/HelloNDKGDB.cpp:6
6               return x;
(gdb) p i
$1 = 10
(gdb) p j
$2 = 20
(gdb) p x
$3 = 200
(gdb) c
Continuing.
```

## How it works...

Along with Android NDK comes a shell script named as `ndk-gdb` to launch a native debugging session with the native code. In order to use `ndk-gdb`, we must build the native code in the debug mode. This will produce a `gdbserver` binary and a `gdb.setup` file along with the native library. At installation, `gdbserver` will be installed and `ndk-gdb` will start `gdbserver` on the Android device.

By default, `ndk-gdb` searches for a running application and attaches `gdbserver` to it. There also are options to launch the application automatically before starting the debugging. Because the application starts first before `gdbserver` is attached to it, some code will be executed before debugging. If we want to debug the code that is executed at the application start up, we can insert a `while(true)` block. After the debugging session starts, we change the flag value to escape from the `while(true)` block. This is demonstrated in our sample project.

Once the debug session starts, we can use `gdb` commands to debug our code.

# Debugging an Android NDK application with CGDB

CGDB is a terminal-based lightweight interface to the GNU debugger `gdb`. It provides a split screen view, which displays the source code along with the debug information. This recipe discusses how to debug Android application with `cgdb`.

## Getting ready

The following instructions install `cgdb` on different operating systems:

▸ If you're using Ubuntu, you can use the following command to install `cgdb`:

```
$ sudo apt-get install cgdb
```

Alternatively, you can download the source code from `http://cgdb.github.com/`, and perform the following instructions to install `cgdb`:

```
$ ./configure --prefix=/usr/local
```

```
$ make
```

```
$ sudo make install
```

Note that `cgdb` requires `libreadline` and `ncurses` development libraries.

▸ If you're using Windows, a Windows binary is available at `http://cgdb.sourceforge.net/download.php`.

▶   If you're using MacOS, you can use the MacPorts installation command as follows:

```
$ sudo port install cgdb
```

Please read the *Debugging Android NDK Application with NDK GDB* recipe before going through this one.

## How to do it...

The following steps enable `cgdb` for Android NDK application debugging:

1.  Make a copy of the `ndk-gdb` script under the Android NDK `root` directory. This can be done with the following command:

    ```
    $ cp $ANDROID_NDK/ndk-gdb $ANDROID_NDK/ndk-cgdb
    ```

    Here, `$ANDROID_NDK` refers to the Android NDK `root` directory.

2.  Change the following line in the `ndk-cgdb` script from:

    ```
    GDBCLIENT=${TOOLCHAIN_PREFIX}gdb
    ```

    To the following:

    ```
    GDBCLIENT="cgdb -d ${TOOLCHAIN_PREFIX}gdb --"
    ```

3.  We'll use the project created in the *Debugging Android NDK application with NDK GDB* recipe. If you don't have the project open in your Eclipse IDE, click on **File | Import**. Select **Existing Projects into Workspace** under **General**, then click on **Next**. In the import window, check **Select root directory**, and browse to the `HelloNDKGDB` project. Click on **Finish** to import the project:

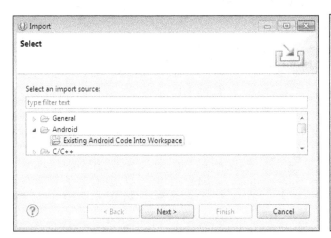

4. Run the application on an Android device. Then, start a termina, and enter the following command:

```
ndk-cgdb
```

The following is a screenshot of the `cgdb` interface:

```
File  Edit  View  Terminal  Help
 1| #include <jni.h>
 2| #include <unistd.h>
 3|
 4| int multiply(int i, int j) {
 5|         int x = i * j;
 6|         return x;
 7| }
 8|
 9| extern "C" {
10|         JNIEXPORT jint JNICALL Java_cookbook_chapter3_HelloNDKGDBActivity_mu
11|         int i = 1, j = 0;
12|         while (i) {
13 -->            j=(++j)/100;
14|         }
15|         return multiply(pi, pj);
16|     }
17| }
/home/roman10/Desktop/book-code/HelloNDKGDB/jni/HelloNDKGDB.cpp
There is NO WARRANTY, to the extent permitted by law.  Type "show copying"
and "show warranty" for details.
This GDB was configured as "--host=x86_64-linux-gnu --target=arm-linux-android".
For bug reporting instructions, please see:
<http://www.gnu.org/software/gdb/bugs/>.
warning: while parsing target library list (at line 2): No segment defined for /
system/bin/linker
BFD: /home/roman10/Desktop/book-code/HelloNDKGDB/obj/local/armeabi/linker: warni
ng: sh_link not set for section `.ARM.exidx'
warning: Could not load shared library symbols for 54 libraries, e.g. libstdc++.
so.
Use the "info sharedlibrary" command to see the complete listing.
Do you need "set solib-search-path" or "set sysroot"?
0x80400dce in Java_cookbook_chapter3_HelloNDKGDBActivity_multiply (env=0xabe0, o
=0x40519970, pi=10, pj=20) at jni/HelloNDKGDB.cpp:13
(gdb)
```

5. We can issue `gdb` commands. Note that the upper-half of the window will mark the current execution line with an arrow and all the breakpoints with red.

## How it works...

As shown in the preceding screenshot, `cgdb` provides a more intuitive interface for debugging the native code in Android. We can view the source code as we enter `gdb` commands. This recipe demonstrates the basic setup of `cgdb` for debugging the native code. The details of how to use `cgdb` can be found at its documentation available at `http://cgdb.github.com/docs/cgdb.html`.

# Debugging an Android NDK application in Eclipse

Debugging at terminals with GDB or CGDB is cumbersome for developers who are used to the graphical development tools. With **Android Development Tools** (**ADT**) 20.0.0 or above, debugging NDK application in Eclipse is fairly easy.

## Getting ready

Make sure you have ADT 20.0.0 or above installed. If not, please refer to recipes in *Chapter 1, Hello NDK*, that explain how to set your environment up.

Make sure you have configured the NDK path in Eclipse. In addition, you are expected to have built and run at least one Android NDK application before reading this recipe. If not, please go through the *Building Android NDK Application at Eclipse* recipe.

## How to do it...

The following steps create a sample Android project and debug it using Eclipse:

1. We'll use the project created in the *Building Android NDK application at Eclipse* recipe. If you don't have the project open in your Eclipse IDE, click on **File** | **Import**. Select **Existing Projects into Workspace** under **General**, then click on **Next**. In the **import** window, check **Select root directory**, and browse to the `HelloNDKEclipse` project. Click on **Finish** to import the project:

2. Right-click on the `HelloNDKEclipse` project, and select **Properties**. In the **Properties** window, select **C/C++ Build**. Uncheck **Use default build command**, and change the **Build** command to `ndk-build NDK_DEBUG=1`.

3. Click on **OK** to dismiss the window:

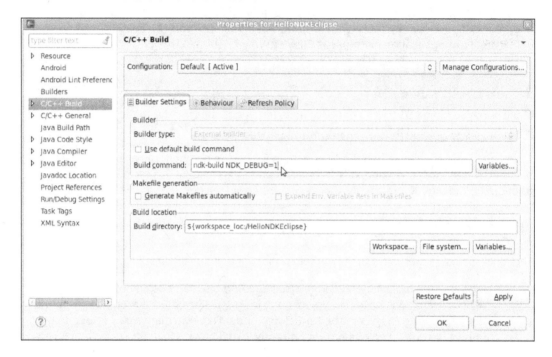

4. Add the following code before the native method is called at `HelloNDKEclipseActivity.java`.

   Set two breakpoints in `HelloNDKEclipse.cpp`:

```
 1 #include <jni.h>
 2
 3 jstring getString(JNIEnv* env) {
 4     return env->NewStringUTF("Hello NDK");
 5 }
 6
 7 extern "C" {
 8     JNIEXPORT jstring JNICALL Java_cookbook_ch
 9         return getString(env);
10     }
11 }
12
```

5. Right-click on your project and then select **Debug As | Android Native Application**. We'll see if the breakpoints are hit.

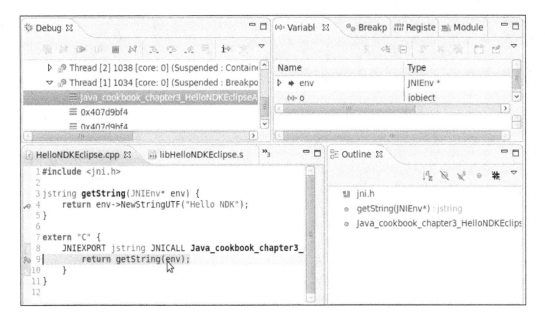

## How it works...

Because there is a delay of a few seconds between the application start and the debugging session start, the source code where the breakpoint is set may have already been executed before debugging. In this case, the breakpoint is never hit. We demonstrated using a `while(true)` loop to overcome this issue in the *Debugging Android NDK application with NDK GDB* recipe. We show another approach here, which sends code to sleep for several seconds at application start. This gives the debugger enough time to start. Once the debugging starts, we can use the normal Eclipse debugging interface to debug our code.

## There's more...

There're a few more debuggers available for debugging Android NDK applications.

**Data Display Debugger** (**DDD**) is a graphical front end for GDB. It is possible to set up DDD to debug Android applications. Detailed instructions for the same can be found at `http://omappedia.org/wiki/Android_Debugging#Debugging_with_GDB_and_DDD`.

**NVIDIA Debug Manager** is an Eclipse plugin that assists in debugging Android NDK applications on devices based on NVIDIA's Tegra platform. More information about this tool can be found at `https://developer.nvidia.com/nvidia-debug-manager-android-ndk`.

# 4
# Android NDK
# OpenGL ES API

In this chapter we will cover the following recipes:

- ▸ Drawing 2D Graphics and applying transforms with OpenGL ES 1.x API
- ▸ Drawing 3D graphics and lighting up the scene with the OpenGL ES 1.x API
- ▸ Mapping texture to 3D objects with the OpenGL ES 1.x API
- ▸ Drawing 3D graphics with the OpenGL ES 2.0 API
- ▸ Displaying graphics with EGL

## Introduction

**Open Graphics Library** (**OpenGL**) is a cross-platform industry standard API for producing 2D and 3D graphics. It specifies a language-independent software interface for graphics hardware or software graphics engines. **OpenGL ES** is a flavor of OpenGL for embedded devices. It consists of a subset of OpenGL specifications and some additional extensions that are specific to OpenGL ES .

OpenGL ES does not require dedicated graphics hardware to work. Different devices can come with graphics hardware with different processing capabilities. The workload of the OpenGL ES calls is divided between the CPU and graphics hardware. It is possible to support OpenGL ES entirely from the CPU. However, graphics hardware can improve performance at different levels, based on its processing capabilities.

Before we dive into Android NDK OpenGL ES, a little introduction of the **Graphics Rendering Pipeline** (**GRP**) in the OpenGL context is necessary. GRP refers to a series of processing stages, which the graphics hardware takes to produce graphics. It accepts objects description in terms of vertices of primitives (**primitives** refer to simple geometric shapes such as point, line, and triangle) and output color values for the pixels on the display. It can be roughly divided into the following four main stages:

1.  **Vertex processing**: It accepts the graphics model description, processes and transforms the individual vertices to project then onto the screen, and combines their information for further processing of **primitives**.

2.  **Rasterization**: It converts primitives into fragments. A **fragment** contains the data that is necessary to generate a pixel's data in the frame buffer. Note that only the pixels affected by one or more primitives will have a fragment. A fragment contains information, such as raster position, depth, interpolated color, and texture coordinates.

3.  **Fragment processing**: It processes each fragment. A series of operations are applied to each fragment, including alpha test, texture mapping, and so on.

4.  **Output merging**: It combines all fragments to produce the color values (including alpha) for the 2D display.

In the modern computer graphics hardware, vertex processing and fragment processing are programmable. We can write programs to perform custom transform and processing of vertices and fragments. In contrast, rasterization and output merging are configurable, but not programmable.

Each of the preceding stages can consist of one or more steps. OpenGL ES 1.x and OpenGL ES 2.0 provide different GRPs. Specifically, OpenGL ES 1.x provides a fixed function pipeline, where we input primitive and texture data, set up lighting, and OpenGL ES will handle the rest. In contrast, OpenGL ES 2.0 provides a programmable pipeline, which allows us to write vertex and fragment shaders in **OpenGL ES Shading Language** (**GLSL**) to handle the specifics.

The following diagram indicaties the fixed function pipeline of OpenGL ES 1.x:

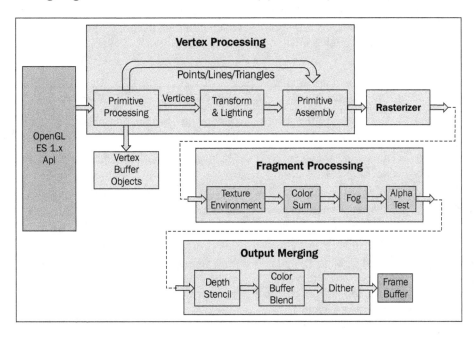

The following is another diagram that illustrates the programmable pipeline of OpenGL ES 2.0:

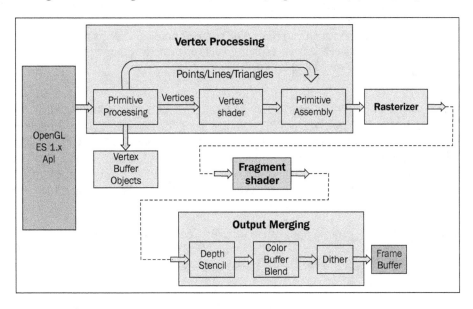

As shown in the preceding diagram, the fixed pipeline in OpenGL ES 1.x has been replaced by the programmable shaders in OpenGL ES 2.0.

With this introduction of computer graphics, we're now ready to start our journey to Android NDK OpenGL ES programming. Android NDK provides both OpenGL ES 1.x (version 1.0 and version 1.1) and OpenGL ES 2.0 libraries, which differ significantly. The following table summarizes the factors to consider when considering the OpenGL ES version to use in our Android applications:

| | OpenGL 1.x | OpenGL 2.0 |
|---|---|---|
| **Performance** | Fast 2D and 3D graphics. | Depending upon the Android device, but in general it provides faster 2D and 3D graphics. |
| **Device compatibility** | Almost all Android devices. | Majority of Android devices, and increasing. |
| **Coding convenience** | Fixed pipeline with convenient functions. Easy to use for simple 3D applications. | No built-in basic functions and more effort may be required for simple 3-D applications. |
| **Graphics control** | Fixed pipeline. Difficult or impossible to create some effects (for example, cartoon shading). | Programmable pipeline. More direct control of the graphics processing pipeline to create certain effects. |

OpenGL ES 1.0 is supported on all Android devices because Android comes with a 1.0-capable software graphics engine, which can be used on devices without corresponding graphics hardware. OpenGL ES 1.1 and OpenGL ES 2.0 are supported only on devices with corresponding **Graphics Processing Unit** (**GPU**).

This chapter will cover both the OpenGL 1.x and OpenGL ES 2.0 APIs in Android NDK. We first demonstrated how to draw 2D and 3D graphics using the OpenGL 1.x API. Transforms, lighting, and texture mapping are also covered. We then introduce the OpenGL 2.0 API in NDK. Lastly, we describe how to display graphics with EGL. This chapter introduces a few basics of computer graphics and principles of OpenGL. Readers who are already familiar with OpenGL ES can skip those parts and focus on how to invoke the OpenGL ES API from Android NDK.

We will provide a sample Android application for every recipe covered in this chapter. Due to space constraints, we cannot show all the source code in the book. Readers are strongly recommended to download the code and refer to to it when going through this chapter.

# Drawing 2D Graphics and applying transforms with the OpenGL ES 1.x API

This recipe covers 2D drawing in OpenGL ES 1.x by example. In order to draw 2D objects, we'll also describe the OpenGL rendering display through `GLSurfaceView`, adding colors to them, and transformation.

## Getting ready

Readers are recommended to read the introduction of this chapter, which is essential to understand some of the content in this recipe.

## How to do it...

The following steps create our sample Android NDK project:

1.  Create an Android application named `TwoDG1`. Set the package name as `cookbook.chapter4.gl1x`. Please refer to the *Loading native libraries and registering native methods* recipe in *Chapter 2, Java Native Interface*, if you want more detailed instructions.

2.  Right-click on the `TwoDG1` project in Eclipse, select **Android Tools | Add Native Support**.

3.  Add the following three Java files under the `cookbook.chapter4.gl1x` package:

    ❏   `MyActivity.java`: It creates the activity of this project:

    ```
    import android.opengl.GLSurfaceView;
    ......
    public class MyActivity extends Activity {
      private GLSurfaceView mGLView;
      @Override
      public void onCreate(Bundle savedInstanceState) {
        super.onCreate(savedInstanceState);
        mGLView = new MySurfaceView(this);
              setContentView(mGLView);
      }
    }
    ```

- ❑ `MySurfaceView.java`: It extends `GLSurfaceView`, which provides a dedicated surface for displaying OpenGL rendering:

```java
public class MySurfaceView extends GLSurfaceView {
  private MyRenderer mRenderer;
  public MySurfaceView(Context context) {
    super(context);
    mRenderer = new MyRenderer();
    this.setRenderer(mRenderer);
    this.setRenderMode(GLSurfaceView.RENDERMODE_WHEN_DIRTY);
  }
}
```

- ❑ `MyRenderer.java`: It implements `Renderer` and calls the native methods:

```java
public class MyRenderer implements GLSurfaceView.Renderer{
  @Override
  public void onSurfaceCreated(GL10 gl, EGLConfig config) {
    naInitGL1x();
  }
  @Override
  public void onDrawFrame(GL10 gl) {
    naDrawGraphics();
  }
  @Override
  public void onSurfaceChanged(GL10 gl, int width, int height) {
    naSurfaceChanged(width, height);
  }
  ......
}
```

4. Add the `TwoDG1.cpp`, `Triangle.cpp`, `Square.cpp`, `Triangle.h`, and `Square.h` files under the `jni` folder. Please refer to the downloaded project for the complete content. Here, we only list some important parts of the code:

`TwoDG1.cpp`: It consists of the code to set up the OpenGL ES 1.x environment and perform the transforms:

```cpp
void naInitGL1x(JNIEnv* env, jclass clazz) {
  glDisable(GL_DITHER);
  glHint(GL_PERSPECTIVE_CORRECTION_HINT, GL_FASTEST);
  glClearColor(0.0f, 0.0f, 0.0f, 1.0f);       glShadeModel(GL_
SMOOTH);       }

void naSurfaceChanged(JNIEnv* env, jclass clazz, int width, int
height) {
  glViewport(0, 0, width, height);
```

```
    float ratio = (float) width / (float)height;
    glMatrixMode(GL_PROJECTION);
    glLoadIdentity();
    glOrthof(-ratio, ratio, -1, 1, 0, 1);  }

void naDrawGraphics(JNIEnv* env, jclass clazz) {
  glClear(GL_COLOR_BUFFER_BIT);
  glMatrixMode(GL_MODELVIEW);
  glLoadIdentity();
  glTranslatef(0.3f, 0.0f, 0.0f);     //move to the right
  glScalef(0.2f, 0.2f, 0.2f);         // Scale down
  mTriangle.draw();
  glLoadIdentity();
  glTranslatef(-0.3f, 0.0f, 0.0f);    //move to the left
  glScalef(0.2f, 0.2f, 0.2f);         // Scale down
glRotatef(45.0, 0.0, 0.0, 1.0);  //rotate
  mSquare.draw();
}
```

`Triangle.cpp`: It draws a 2D triangle:

```
void Triangle::draw() {
  glEnableClientState(GL_VERTEX_ARRAY);
  glVertexPointer(3, GL_FLOAT, 0, vertices);
  glColor4f(0.5f, 0.5f, 0.5f, 0.5f);      //set the current color
  glDrawArrays(GL_TRIANGLES, 0, 9/3);
  glDisableClientState(GL_VERTEX_ARRAY);
}
```

`Square.cpp`: It draws a 2D square:

```
void Square::draw() {
  glEnableClientState(GL_VERTEX_ARRAY);
  glEnableClientState(GL_COLOR_ARRAY);
  glVertexPointer(3, GL_FLOAT, 0, vertices);
  glColorPointer(4, GL_FLOAT, 0, colors);
  glDrawElements(GL_TRIANGLES, 6, GL_UNSIGNED_BYTE, indices);
  glDisableClientState(GL_VERTEX_ARRAY);
  glDisableClientState(GL_COLOR_ARRAY);
}
```

5. Add the `Android.mk` file under the `jni` folder with following content:

```
LOCAL_PATH := $(call my-dir)
include $(CLEAR_VARS)
LOCAL_MODULE    := TwoDG1
LOCAL_SRC_FILES := Triangle.cpp Square.cpp TwoDG1.cpp
LOCAL_LDLIBS := -lGLESv1_CM -llog
include $(BUILD_SHARED_LIBRARY)
```

6. Build the Android NDK application and run it on an Android device. The following is a screenshot of the display:

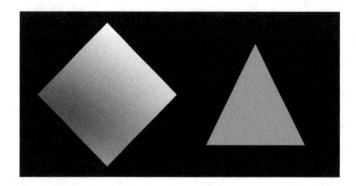

## How it works...

This recipe demonstrates basic 2D drawing with OpenGL ES.

### OpenGL ES rendering display through GLSurfaceView

GLSurfaceView and GLSurfaceView.Renderer are the two foundational classes provided by Android SDK to display OpenGL ES graphics.

GLSurfaceView accepts a user defined Renderer object that does the actual rendering. It is often extended to handle touch events, which is illustrated in the next recipe. It supports both on-demand and continuous rendering. In our sample code, we simply set the Renderer object and configure the rendering mode to on-demand.

GLSurfaceView.Renderer is the interface for renderer. Three methods need to be implemented with it:

 ▶ onSurfaceCreated: It's called once when setting up the OpenGL ES environment.

 ▶ onSurfaceChanged: It's called if the geometry of the view changes; most common examples are device screen orientation changes.

 ▶ onDrawFrame: It's called at each redraw of the view.

In our sample project, MyRenderer.java is a simple wrapper, while the actual work is done in native C++ code.

### Drawing objects at OpenGL ES

Two methods are commonly used to draw objects in OpenGL ES, including glDrawArrays and glDrawElements. We demonstrate the usage of these two methods in Triangle.cpp and Square.cpp respectively. Note that both the methods require GL_VERTEX_ARRAY to be enabled.

The first argument is the mode of drawing, which indicates the primitive to use. In our sample code, we used GL_TRIANGLES, which means we're actually drawing two triangles to form the square. Other valid values in Android NDK OpenGL ES include GL_POINTS, GL_LINES, GL_LINE_LOOP, GL_LINE_STRIP, GL_TRIANGLE_STRIP, and GL_TRIANGLE_FAN.

## Colors at OpenGL ES

We also demonstrate two methods to add colors to the objects. In Triangle.cpp, we set the current color by the glColor4f API call. In Square.cpp, we enable GL_COLOR_ARRAY, and define an array of color by using glColorPointer. The array of colors will be used by the glDrawElements (it's also OK to use glDrawArrays) API call.

## OpenGL ES transformation

The following diagram illustrates different transformation stages in OpenGL ES 1.0:

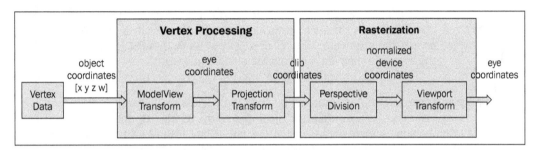

As illustrated in the diagram, vertex data are transformed before rasterization. The transforms are analogous to taking a photograph with a camera:

▸ **ModelView transform**: Arrange the scene and place the camera

▸ **Projection transform**: Choose a camera lens and adjust zoom factor

▸ **Viewpoint transform**: Determine how large the final photograph is

**ModelView transform** actually refers to two different transforms, namely Model Transform and View Transform. **Model Transform** refers to the process of converting all objects from its object space (also known as local space or model space) to a world space, which is shared among all objects. This transform is done through a series of scaling (glScalef), rotation (glRotatef) and translation (glTranslatef).

▸ glScalef: It stretches, shrinks, or reflects an object. The x-, y- and z-axis values are multiplied by the corresponding x, y, and z scaling factor. In our sample code, we called glScalef(0.2f, 0.2f, 0.2f) to scale down both the triangle and the square, so that they can fit into the screen.

- ▸ `glRotatef`: It rotates an object in a counter clockwise manner in the direction from the origin through specified point (x, y, z). The rotation angle is measured in degrees. In our sample code, we called `glRotatef(45.0, 0.0, 0.0, 1.0)` to rotate the square about the z-axis by 45 degrees.

- ▸ `glTranslatef`: It moves an object by the given values along each axis. In our sample code, we called `glTranslatef(0.3f, 0.0f, 0.0f)` to move the triangle to the right and `glTranslatef(-0.3f, 0.0f, 0.0f)` to move the square to the left, so that they won't overlap.

Model transform arranges the objects in a scene, while View transform changes the position of the viewing camera. To produce a specific image, we can either move the objects or change our camera position. Therefore, OpenGL ES internally performs the two transforms using a single matrix – the +`GL_MODELVIEW` matrix.

> OpenGL ES defines that the camera is default at the origin (0, 0, 0) of eye coordinates space and aims into the negative z-axis. It is possible to change the position by `GLU.gluLookAt` at the Android SDK. However, the corresponding API is not available at Android NDK.

**Projection transform** determines what can be seen (analogous to choosing camera lens and zoom factor) and how vertex data are projected onto the screen. OpenGL ES supports two modes of projection, namely perspective projection (`glFrustum`) and orthographic projection (`glOrtho`). **Perspective projection** makes objects that are farther away smaller, which matches with a normal camera. On the other hand, **Orthographic projection** is analogous to the telescope, which maps objects directly without affecting their size. OpenGL ES manipulates the transform through the `GL_PROJECTION` matrix. After a project transform, objects which are outside of the clipping volume are clipped out and not drawn in the final scene. In our sample project, we called `glOrthof(-ratio, ratio, -1, 1, 0, 10)` to specify the viewing volume, where `ratio` refers to the width to height ratio of the screen.

After projection transform, perspective division is done by dividing the clip coordinates by the transformed `w` value of the input vertex. The values the for x-, y-, and z-axes will be normalized to the range between `-1.0` to `1.0`.

The final stage of the OpenGL ES transform pipeline is the viewport transform, which maps the normalized device coordinates to window coordinates (in pixels, with the origin at the upper-left corner). Note that a viewpoint also includes a z component, which is needed for situations, such as ordering of two overlapping OpenGL scenes, and can be set with the `glDepthRange` API call. Applications usually need to set viewport when the display size changes through the `glViewport` API call. In our example, we set the viewport as the entire screen by calling `glViewport(0, 0, width, height)`. This setting, together with the `glOrthof` call, will keep the objects in proportion after projection transform, as shown in the following diagram:

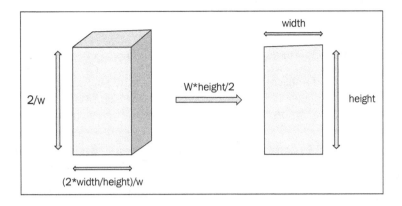

As shown in the diagram, the clipping volume is set to (-width/height, width/height, -1, 1, 0, 1). At perspective division, the vertex is divided by w. At viewpoint transform, both the x and y coordinates ranges are scaled up by `w*height/2`. Therefore, the objects will be in proportion as shown in the *How to do it...* section of this recipe. The left-had side of the following screenshot shows the output, if we set clipping volume by calling `glOrthof(-1, 1, -1, 1, 0, 1)`, and the right one indicates what the graphics look like if viewport is set by calling `glViewport(0, 0, width/2, height/5)`:

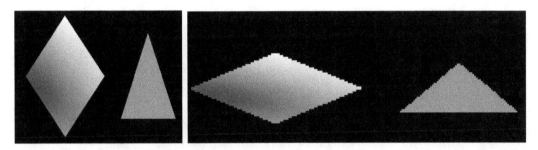

# Drawing 3D graphics and lighting up the scene with the OpenGL ES 1.x API

This recipe covers how to draw 3D objects, handle touch events, and lighten up the objects in OpenGL ES.

## Getting ready

Readers are recommended to read the introduction and the *Drawing 2D Graphics and Apply Transforms with OpenGL ES 1.x API* recipies below before going through this one.

## How to do it...

The following steps show how to develop our sample Android project:

1. Create an Android application named `CubeG1`. Set the package name as `cookbook.chapter4.gl1x`. Please refer to the *Loading native libraries and registering native methods* recipe in *Chapter 2, Java Native Interface*, if you want more detailed instructions.

2. Right-click on the project CubeG1, select **Android Tools | Add Native Support**.

3. Add three Java files, namely `MyActivity.java`, `MySurfaceView`, and `MyRenderer.java`, under the `cookbook.chapter4.gl1x` package. `MyActivity.java` is the same as used in the previous recipe.

`MySurfaceView.java` extends `GLSurfaceView` with the code to handle touch events:

```
public class MySurfaceView extends GLSurfaceView {
  private MyRenderer mRenderer;
  private float mPreviousX;
   private float mPreviousY;
   private final float TOUCH_SCALE_FACTOR = 180.0f / 320;
  public MySurfaceView(Context context) {
    super(context);
    mRenderer = new MyRenderer();
    this.setRenderer(mRenderer);
    //control whether continuously drawing or on-demand
    this.setRenderMode(GLSurfaceView.RENDERMODE_WHEN_DIRTY);
  }

  public boolean onTouchEvent(final MotionEvent event) {
    float x = event.getX();
      float y = event.getY();
      switch (event.getAction()) {
      case MotionEvent.ACTION_MOVE:
          float dx = x - mPreviousX;
          float dy = y - mPreviousY;
          mRenderer.mAngleX += dx * TOUCH_SCALE_FACTOR;
          mRenderer.mAngleY += dy * TOUCH_SCALE_FACTOR;
          requestRender();
      }
      mPreviousX = x;
      mPreviousY = y;
      return true;
  }
}
```

`MyRenderer.java` implements a render to invoke the native methods to render the graphics:

```java
public class MyRenderer implements GLSurfaceView.Renderer{
    public float mAngleX;
    public float mAngleY;
    @Override
    public void onSurfaceCreated(GL10 gl, EGLConfig config) {
        naInitGL1x();
    }
    @Override
    public void onDrawFrame(GL10 gl) {
        naDrawGraphics(mAngleX, mAngleY);
    }
    @Override
    public void onSurfaceChanged(GL10 gl, int width, int height) {
        naSurfaceChanged(width, height);
    }
}
```

4. Add the `CubeG1.cpp`, `Cube.cpp`, and `Cube.h` files under the `jni` folder. Please refer to downloaded project for the complete content. Let's list out the code for the `naInitGL1x`, `naSurfaceChanged`, and `naDrawGraphics` native methods in `CubeG1.cpp`, and draw and lighting methods in `Cube.cpp`:

`CubeG1.cpp` sets up the OpenGL ES environment and lighting:

```cpp
void naInitGL1x(JNIEnv* env, jclass clazz) {
    glDisable(GL_DITHER);
    glHint(GL_PERSPECTIVE_CORRECTION_HINT, GL_NICEST);
    glClearColor(0.0f, 0.0f, 0.0f, 1.0f);      glEnable(GL_CULL_FACE);
    glClearDepthf(1.0f);
    glEnable(GL_DEPTH_TEST);
    glDepthFunc(GL_LEQUAL);      //type of depth test
    glShadeModel(GL_SMOOTH);
    glLightModelx(GL_LIGHT_MODEL_TWO_SIDE, 0);
    float globalAmbientLight[4] = {0.5, 0.5, 0.5, 1.0};
    glLightModelfv(GL_LIGHT_MODEL_AMBIENT, globalAmbientLight);
    GLfloat lightOneDiffuseLight[4] = {1.0, 1.0, 1.0, 1.0};
    GLfloat lightOneSpecularLight[4] = {1.0, 1.0, 1.0, 1.0};
    glLightfv(GL_LIGHT0, GL_DIFFUSE, lightOneDiffuseLight);
    glLightfv(GL_LIGHT0, GL_SPECULAR, lightOneSpecularLight);
    glEnable(GL_LIGHTING);
    glEnable(GL_LIGHT0);
}
void naSurfaceChanged(JNIEnv* env, jclass clazz, int width, int
height) {
```

```
    glViewport(0, 0, width, height);
     float ratio = (float) width / height;
    glMatrixMode(GL_PROJECTION);
    glLoadIdentity();
    glOrthof(-ratio, ratio, -1, 1, -10, 10);
}
void naDrawGraphics(JNIEnv* env, jclass clazz, float pAngleX,
float pAngleY) {
  glClear(GL_COLOR_BUFFER_BIT | GL_DEPTH_BUFFER_BIT);
   glMatrixMode(GL_MODELVIEW);
   glLoadIdentity();
   glRotatef(pAngleX, 0, 1, 0);  //rotate around y-axis
   glRotatef(pAngleY, 1, 0, 0);  //rotate around x-axis
  glScalef(0.3f, 0.3f, 0.3f);       // Scale down
mCube.lighting();
  mCube.draw();
  float lightOnePosition[4] = {0.0, 0.0, 1.0, 0.0};
  glLightfv(GL_LIGHT0, GL_POSITION, lightOnePosition);
}
```

Cube.cpp draws a 3D cube and lightens it up:

```
void Cube::draw() {
  glEnableClientState(GL_VERTEX_ARRAY);
  glVertexPointer(3, GL_FLOAT, 0, vertices);
  glDrawElements(GL_TRIANGLES, 36, GL_UNSIGNED_BYTE, indices);
  glDisableClientState(GL_VERTEX_ARRAY);
}
void Cube::lighting() {
  GLfloat cubeOneAmbientFraction[4] = {0.0, 0.5, 0.5, 1.0};
  GLfloat cubeOneDiffuseFraction[4] = {0.8, 0.0, 0.0, 1.0};
  GLfloat cubeSpecularFraction[4] = {0.0, 0.0, 0.0, 1.0};
  GLfloat cubeEmissionFraction[4] = {0.0, 0.0, 0.0, 1.0};
  glMaterialfv(GL_FRONT_AND_BACK, GL_AMBIENT,
cubeOneAmbientFraction);
  glMaterialfv(GL_FRONT_AND_BACK, GL_DIFFUSE,
cubeOneDiffuseFraction);
  glMaterialfv(GL_FRONT_AND_BACK, GL_SPECULAR,
cubeSpecularFraction);
  glMaterialfv(GL_FRONT_AND_BACK, GL_EMISSION,
cubeEmissionFraction);
  glMaterialf(GL_FRONT_AND_BACK, GL_SHININESS, 60.0);
}
```

5. Add the `Android.mk` file under the `jni` folder with the following content:

```
LOCAL_PATH := $(call my-dir)
include $(CLEAR_VARS)
LOCAL_MODULE     := CubeG1
LOCAL_SRC_FILES := Cube.cpp CubeG1.cpp
LOCAL_LDLIBS := -lGLESv1_CM -llog
include $(BUILD_SHARED_LIBRARY)
```

6. Build the Android NDK application and run it on an Android device. The app will display a cube, which we can touch to rotate:

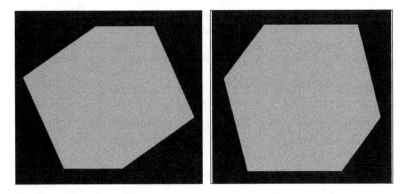

## How it works...

This recipe discusses how to use the OpenGL ES 1.x API to draw 3D graphics. Note that we will need to load the OpenGL ES library GLESv1_CM in the `Andorid.mk` file, and include the header file `GLES/gl.h` in the native source code.

▶ **Drawing 3D objects in OpenGL ES**: Drawing 3D objects is similar to drawing 2D objects. In the `Cube::draw` method, we first set up the vertex buffer and then called `glDrawElements` to draw the six faces of the cube. We used GL_TRIANGLES as a primitive. Since each face contains two triangles, there are 12 triangles and 36 vertices.

▶ **Touch event handling**: In `MySurfaceView.java`, we override the `onTouchEvent` method to detect the figure movement on screen and change the rotation angle attributes of `MyRenderer`. We call the `requestRender` method to request the renderer to redraw the graphics.

> ▸ **Lighting and material in OpenGL ES**: There are classes of lighting models, namely local illumination and global illumination. **Local illumination** only considers direct lighting, and therefore lighting calculation can be performed on individual objects. In contrast to this, **global illumination** takes indirect lighting reflected from other objects and the environment, and therefore is more computation-expensive. Local illumination is used in OpenGL ES 1.x, while the global illumination can be programmed using the **OpenGL Shading Language** (**GLSL**) in OpenGL ES 2.0. Here, we discuss lighting in OpenGL ES 1.x only.

Three parties are involved in OpenGL ES when lighting is considered, including camera position, light sources, and the material of the objects. Camera position is always at a default position (0, 0, 0) and aims into the negative z-axis, as discussed in the previous recipe. Light sources can provide separate ambient, diffuse, and specular lights. Materials can reflect different amounts of ambient, diffuse, and specular lights. In addition, materials may also emit light. Each of the light consists of RGB components:

> ▸ **Ambient light**: It approximates the constant amount of light present everywhere in the scene.

> ▸ **Diffuse light**: It approximates the light from distant directional light source (for example, sunlight). When the reflected light strikes a surface, it is scattered equally in all directions.

> ▸ **Specular light**: It approximates the lights reflected by a smooth surface. Its intensity depends on the angle between the viewer and the direction of the ray reflected from the surface.

> ▸ **Emission light**: Some materials can emit lights.

Note that RGB values in light sources indicate the intensity of the color component, while they refer to the reflected proportions of those colors in the material. To understand how both the light sources and material can affect the viewer perception of the object, think of a white light that strikes on a surface, which only reflects blue component of the light, then the surface will appear as blue for the viewer. If the light is pure red, the surface will be black for the viewer.

The following steps can be performed to set up simple lighting in OpenGL ES:

1. Set the lighting model parameters. This is done through `glLightModelfv`. Android NDK OpenGL ES supports two parameters, including `GL_LIGHT_MODEL_AMBIENT` and `GL_LIGHT_MODEL_TWO_SIDE`. The first one allows us to specify the global ambient light, and the second one allows us to specify whether we want to calculate lighting at the back of the surface.

2. Enable, configure, and place one or more light sources. This is done through the `glLightfv` method. We can configure ambient, diffuse, and specular light separately. The light source position is also configured through `glLightfv` with `GL_POSITION`. In `CubeG1.cpp`, we used the following code:

```
float lightOnePosition[4] = {0.0, 0.0, 1.0, 0.0};
glLightfv(GL_LIGHT0, GL_POSITION, lightOnePosition);
```

The fourth value of the position indicates whether the light source is positional or directional. When the value is set to 0, the light is directional, which simulates a light source that is far away (sunlight). The light rays are parallel when hitting the surface, and the (x, y, z) values of the position refer to the direction of the light. If the fourth value is set to 1, the light is positional, which is similar to a light bulb. The (x, y, z) values refer to the position of the light source and the light rays hit the surface from different angles. Note that the light source emits light at equal intensities to all directions. The two kinds of lighting sources are illustrated in the following image:

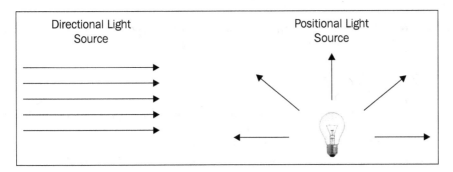

Besides positional and directional light sources, there's also spotlight:

1.  We shall enable lighting and the light sources also by calling

    `glEnable(GL_LIGHTING);`

    and

    `glEnable(GL_LIGHTx);`

2.  Define the normal vectors for each vertex of all objects. The orientation of the object relative to the light sources is determined by these normals. In our code, we rely on OpenGL ES's default normals.

3.  Define the material. This is done by the `glMaterialf` or `glMaterialfv` method. In our sample code, we specify the red component of the diffuse light to be 0.8, while keeping the green and blue components 0. Therefore, the final cube appears to be red.

# Mapping texture to 3D objects with the OpenGL ES 1.x API

**Texture mapping** is a technique that overlays an image onto an object's surface to create a more realistic scene. This recipe covers how to add texture in OpenGL ES 1.x.

## Getting ready

Readers are recommended to read the *Drawing 3D graphics and lighting up the scene with OpenGL ES 1.x API* recipe before going through this one.

## How to do it...

The following steps create an Android project that demonstrates mapping texture to 3D objects:

1. Create an Android application named DiceG1. Set the package name as cookbook.chapter4.gl1x. Please refer to the *Loading native libraries and registering native methods* recipe in *Chapter 2, Java Native Interface*, if you want more detailed instructions.

2. Right-click on the project CubeG1, select **Android Tools | Add Native Support**.

3. Add three Java files, namely MyActivity.java, MySurfaceView.java, and MyRenderer.java under the cookbook.chapter4.diceg1 package. MyActivity.java and MySurfaceView.java are similar to the previous recipe.

4. MyRenderer.java is listed as follows:

```java
public class MyRenderer implements GLSurfaceView.Renderer{
    public float mAngleX;
    public float mAngleY;
    private Context mContext;
    public MyRenderer(Context pContext) {
        super();
        mContext = pContext;
    }
    @Override
    public void onSurfaceCreated(GL10 gl, EGLConfig config) {
        //call native methods to load the textures
        LoadTexture(R.drawable.dice41, mContext, 0);
        LoadTexture(R.drawable.dice42, mContext, 1);
        LoadTexture(R.drawable.dice43, mContext, 2);
        LoadTexture(R.drawable.dice44, mContext, 3);
        LoadTexture(R.drawable.dice45, mContext, 4);
        LoadTexture(R.drawable.dice46, mContext, 5);
        naInitGL1x();
    }
```

```
… …
    private void LoadTexture(int resId, Context context, int texIdx)
{
    //Get the texture from the Android resource directory
    InputStream is = context.getResources().
openRawResource(resId);
    Bitmap bitmap = null;
    try {
        BitmapFactory.Options options = new BitmapFactory.Options();
        options.inPreferredConfig = Bitmap.Config.ARGB_8888;
        bitmap = BitmapFactory.decodeStream(is, null, options);
        naLoadTexture(bitmap, bitmap.getWidth(), bitmap.getHeight(),
texIdx);
    } finally {
        try {
            is.close();
            is = null;
        } catch (IOException e) {
        }
    }
    if (null != bitmap) {
        bitmap.recycle();
    }
}
}
```

5. Add the `DiceG1.cpp`, `Cube.cpp`, `Cube.h`, and `mylog.h` files under the `jni` folder. Please refer to the downloaded project for the complete content. Here, we list out the code the for`naLoadTexture` and `naInitGL1x` native methods in `DiceG1.cpp`, and the `draw` method in `Cube.cpp`:

```
void naLoadTexture(JNIEnv* env, jclass clazz, jobject pBitmap, int
pWidth, int pHeight, int pId) {
    int lRet;
    AndroidBitmapInfo lInfo;
    void* l_Bitmap;
    GLint format;
    GLenum type;
    if ((lRet = AndroidBitmap_getInfo(env, pBitmap, &lInfo)) < 0) {
        return;
    }
    if (lInfo.format == ANDROID_BITMAP_FORMAT_RGB_565) {
        format = GL_RGB;
        type = GL_UNSIGNED_SHORT_5_6_5;
    } else if (lInfo.format == ANDROID_BITMAP_FORMAT_RGBA_8888) {
        format = GL_RGBA;
```

```
      type = GL_UNSIGNED_BYTE;
    } else {
    return;
    }
    if ((lRet = AndroidBitmap_lockPixels(env, pBitmap, &l_Bitmap)) <
0) {
      return;
    }
    glGenTextures(1, &texIds[pId]);
    glBindTexture(GL_TEXTURE_2D, texIds[pId]);
    glTexParameteri(GL_TEXTURE_2D, GL_TEXTURE_MIN_FILTER, GL_
NEAREST);
    glTexParameteri(GL_TEXTURE_2D, GL_TEXTURE_MAG_FILTER, GL_
NEAREST);
    glTexParameteri(GL_TEXTURE_2D, GL_TEXTURE_WRAP_S, GL_REPEAT);
    glTexParameteri(GL_TEXTURE_2D, GL_TEXTURE_WRAP_T, GL_REPEAT);
    glTexImage2D(GL_TEXTURE_2D, 0, format, pWidth, pHeight, 0,
format, type, l_Bitmap);
    AndroidBitmap_unlockPixels(env, pBitmap);
}
void naInitGL1x(JNIEnv* env, jclass clazz) {
  glDisable(GL_DITHER);
  glHint(GL_PERSPECTIVE_CORRECTION_HINT, GL_NICEST);
  glClearColor(0.0f, 0.0f, 0.0f, 1.0f);
  glEnable(GL_CULL_FACE);
  glClearDepthf(1.0f);
  glEnable(GL_DEPTH_TEST);
  glDepthFunc(GL_LEQUAL);
  glShadeModel(GL_SMOOTH);
  mCube.setTexCoords(texIds);
  glTexEnvx(GL_TEXTURE_ENV, GL_TEXTURE_ENV_MODE, GL_REPLACE);
  glEnable(GL_TEXTURE_2D);
}
Cube.cpp: drawing the cube and mapping texture
void Cube::draw() {
  glEnableClientState(GL_VERTEX_ARRAY);
  glEnableClientState(GL_TEXTURE_COORD_ARRAY);  // Enable texture-
coords-array
  glFrontFace(GL_CW);

  glBindTexture(GL_TEXTURE_2D, texIds[0]);
  glTexCoordPointer(2, GL_FLOAT, 0, texCoords);
  glVertexPointer(3, GL_FLOAT, 0, vertices);
  glDrawElements(GL_TRIANGLES, 18, GL_UNSIGNED_BYTE, indices);

….
  glDisableClientState(GL_VERTEX_ARRAY);
  glDisableClientState(GL_TEXTURE_COORD_ARRAY);
}
```

6. Add the `Android.mk` file under the `jni` folder with the following content:

```
LOCAL_PATH := $(call my-dir)
include $(CLEAR_VARS)
LOCAL_MODULE     := DiceG1
LOCAL_SRC_FILES := Cube.cpp DiceG1.cpp
LOCAL_LDLIBS := -lGLESv1_CM -llog -ljnigraphics
include $(BUILD_SHARED_LIBRARY)
```

7. Build the Android NDK application and run it on an Android device. The app will display a cube textured as a dice:

## How it works...

This recipe adds a texture to the 3D cube to make it look like a dice.

▸ **Texture coordinates**: A texture is typically a 2D image. Texture coordinates (`s`, `t`) are usually normalized to [`0.0`, `1.0`] as shown in the following diagram. Texture image is mapped to [`0`, `1`] in both the `s` and `t` axes:

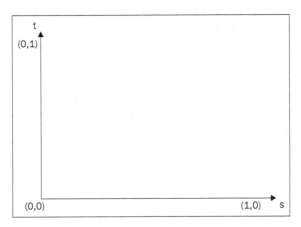

- ▶ **Loading textures**: The first step of mapping texture in OpenGL ES is to load them. In our example, we used Android SDK to read image files from drawable resources and pass the bitmaps to native code. The native method `naLoadTexture` locks the bitmap image and performs the following OpenGL operations.

    - ❑ **Create the glGenTexture texture**: This generates texture IDs.

    - ❑ Bind texture: glBindTexture. This tells OpenGL which texture id we're working with.

    - ❑ **Set the texture filtering**: `glTexParameter` with `GL_TEXTURE_MIN_FILTER` or `GL_TEXTURE_MAG_FILTER` (this is discussed later).

    - ❑ **Set the texture wrapping**: `glTexParameter` with `GL_TEXTURE_WRAP_S` or `GL_TEXTURE_WRAP_T` (this is discussed later).

    - ❑ **Load the images data to OpenGL**: (`glTexImage2D`) we need to specify image data, width, height, color format, and so on.

- ▶ **Texture wrapping**: texture is mapped to `[0, 1]` in both the s and t axes. However, we can specify the texture coordinates beyond the range. Wrapping will be applied once that happens. Typical settings for texture wrapping are as follows:

    - ❑ `GL_CLAMP`: Clamp the texture coordinates to `[0.0, 1.0]`.

    - ❑ `GL_REPEAT`: Repeat the texture. This creates a repeating pattern.

- ▶ **Texture filtering**: It is common that the texture image has a different resolution than the object. If the texture is smaller, magnification is performed; if the texture is larger, minification is performed. The following two methods are used generally:

    - ❑ `GL_NEAREST`: Use the texture element that is nearest to the center of the pixel being textured.

    - ❑ `GL_LINEAR`: Apply interpolation to calculate the color values based on the four texture elements closest to the pixel being textured.

- ▶ **Set the texture environment**: Before we map textures to objects, we can call `glTexEnvf` to control how texture values are interpreted when a fragment is textured. We can configure `GL_TEXTURE_ENV_COLOR` and `GL_TEXTURE_ENV_MODE`. In our sample project, we used the `GL_REPLACE` for `GL_TEXTURE_ENV_MODE`, which simply replaces the cube fragments with texture values.

- ▶ **Mapping the texture**: We draw each face of the 3D cube and map the texture by `glDrawElement`. `GL_TEXTURE_COORD_ARRAY` must be enabled by calling `glEnableClientState`. Before drawing each interface, we bind to the corresponding texture by calling `glBindTexture`.

In our native code, we used the Android native bitmap API to receive texture bitmap object from Java code. More details of this API will be covered in *Chapter 7, Other Android NDK API*.

# Drawing 3D graphics with the OpenGL ES 2.0 API

The previous recipes describe OpenGL ES 1.x on the Android NDK. This recipe covers how to use OpenGL ES 2.0 in Android NDK.

## Getting ready

Readers are recommended to read the introduction of this chapter before going through this recipe. A lot of graphic basics are covered in the following recipes; it is suggested that we go through them first:

- *Drawing 2D graphics and applying transforms with OpenGL ES 1.x API*
- *Drawing 3D graphics and lighting up the scene with OpenGL ES 1.x API*

## How to do it...

The following steps create an Android project that renders a 3D cube with OpenGL ES 2.0 API in Android NDK:

1. Create an Android application named CubeG2. Set the package name as cookbook.chapter4.cubeg2. Please refer to the *Loading native libraries and registering native methods* recipe of *Chapter 2, Java Native Interface*, if you want more detailed instructions.

2. Right-click on the project CubeG2, select **Android Tools | Add Native Support**.

3. Add three Java files, namely MyActivity.java, MyRenderer.java, and MySurfaceView.java. We only list a part of MyRenderer.java here, since the other two files—MyActivity.java and MySurfaceView.java—are similar to the files in the previous recipe:

```
@Override
public void onSurfaceCreated(GL10 gl, EGLConfig config) {
    String vertexShaderStr = LoadShaderStr(mContext, R.raw.
vshader);
    String fragmentShaderStr = LoadShaderStr(mContext, R.raw.
fshader);
    naInitGL20(vertexShaderStr, fragmentShaderStr);
}
```

```
@Override
public void onDrawFrame(GL10 gl) {
  naDrawGraphics(mAngleX, mAngleY);
}
@Override
public void onSurfaceChanged(GL10 gl, int width, int height) {
  naSurfaceChanged(width, height);
}
```

4. Add the `Cube.cpp`, `matrix.cpp`, `CubeG2.cpp`, `Cube.h`, `matrix.h`, and `mylog.h` files under the `jni` folder. The content of the files are summarized as follows:

  □ **Cube.cpp and Cube.h**: They define a `Cube` object and method to draw a 3D cube.

  □ **matrix.cpp and matrix.h**: These matrix operations, including creating translation, scale and rotation matrices, and matrix multiplication.

  □ **CubeG2.cpp**: They create and load shaders. They also create, link, and use programs and apply transformations to the 3D cube.

  □ **mylog.h**: They define macros for Android NDK logging.

Here, we list a part of `Cube.cpp` and `CubeG2.cpp`.

`Cube.cpp`:

```
...
void Cube::draw(GLuint pvPositionHandle) {
  glVertexAttribPointer(pvPositionHandle, 3, GL_FLOAT, GL_FALSE,
0, vertices);
  glEnableVertexAttribArray(pvPositionHandle);
  glDrawArrays(GL_TRIANGLES, 0, 36);
}
...
```

`CubeG2.cpp`: It includes the `loadShader`, `createProgram`, `naInitGL20`, and `naDrawGraphics` methods, which are explained as follows:

  □ `loadShader`: This method creates a shader, attaches a source, and compiles the shader:

```
GLuint loadShader(GLenum shaderType, const char* pSource) {
    GLuint shader = glCreateShader(shaderType);
    if (shader) {
        glShaderSource(shader, 1, &pSource, NULL);
        glCompileShader(shader);
        GLint compiled = 0;
        glGetShaderiv(shader, GL_COMPILE_STATUS, &compiled);
        if (!compiled) {
```

```
            GLint infoLen = 0;
            glGetShaderiv(shader, GL_INFO_LOG_LENGTH,
    &infoLen);
            if (infoLen) {
                char* buf = (char*) malloc(infoLen);
                if (buf) {
                    glGetShaderInfoLog(shader, infoLen, NULL,
    buf);
                    free(buf);
                }
                glDeleteShader(shader);
                shader = 0;
            }
        }
    }
    return shader;
}
```

- ❏ `createProgram`: This method creates a program object, attaches shaders, and links the program:

```
GLuint createProgram(const char* pVertexSource, const char*
pFragmentSource) {
    GLuint vertexShader = loadShader(GL_VERTEX_SHADER,
pVertexSource);
    GLuint pixelShader = loadShader(GL_FRAGMENT_SHADER,
pFragmentSource);
    GLuint program = glCreateProgram();
    if (program) {
        glAttachShader(program, vertexShader);
        glAttachShader(program, pixelShader);
        glLinkProgram(program);
    }
    return program;
}
```

- ❏ `naInitGL20`: This method sets up the OpenGL ES 2.0 environment, gets the shader source string, and gets the shader attribute and uniform positions:

```
void naInitGL20(JNIEnv* env, jclass clazz, jstring
vertexShaderStr, jstring fragmentShaderStr) {
  glDisable(GL_DITHER);
  glClearColor(0.0f, 0.0f, 0.0f, 1.0f);
glClearDepthf(1.0f);
  glEnable(GL_DEPTH_TEST);
  glDepthFunc(GL_LEQUAL);
    const char *vertexStr, *fragmentStr;
  vertexStr = env->GetStringUTFChars(vertexShaderStr, NULL);
  fragmentStr = env->GetStringUTFChars(fragmentShaderStr,
NULL);
```

```
    setupShaders(vertexStr, fragmentStr);
    env->ReleaseStringUTFChars(vertexShaderStr, vertexStr);
    env->ReleaseStringUTFChars(fragmentShaderStr,
fragmentStr);
    gvPositionHandle = glGetAttribLocation(gProgram,
"vPosition");
    gmvP = glGetUniformLocation(gProgram, "mvp");

}
```

❑ `naDrawGraphics`: This method applies model transforms (rotate, scale, and translate) and the projection transform:

```
void naDrawGraphics(JNIEnv* env, jclass clazz, float
pAngleX, float pAngleY) {
    glClear(GL_COLOR_BUFFER_BIT | GL_DEPTH_BUFFER_BIT);
    glClearColor(0.0, 0.0, 0.0, 1.0f);
    glUseProgram(gProgram);
//  GL1x: glRotatef(pAngleX, 0, 1, 0);  //rotate around
y-axis
//  GL1x: glRotatef(pAngleY, 1, 0, 0);  //rotate around
x-axis
    //rotate
    rotate_matrix(pAngleX, 0.0, 1.0, 0.0, aRotate);
    rotate_matrix(pAngleY, 1.0, 0.0, 0.0, aModelView);
    multiply_matrix(aRotate, aModelView, aModelView);
//  GL1x: glScalef(0.3f, 0.3f, 0.3f);        // Scale down
    scale_matrix(0.5, 0.5, 0.5, aScale);
    multiply_matrix(aScale, aModelView, aModelView);
// GL1x: glTranslate(0.0f, 0.0f, -3.5f);
    translate_matrix(0.0f, 0.0f, -3.5f, aTranslate);
    multiply_matrix(aTranslate, aModelView, aModelView);
//  gluPerspective(45, aspect, 0.1, 100);
    perspective_matrix(45.0, (float)gWidth/(float)gHeight,
0.1, 100.0, aPerspective);
    multiply_matrix(aPerspective, aModelView, aMVP);
    glUniformMatrix4fv(gmvP, 1, GL_FALSE, aMVP);
    mCube.draw(gvPositionHandle);
}
```

5. Create a folder named `raw` under the `res` folder, and add the following two files to it:

❑ `vshader`: This is the vertex shader source:

```
attribute vec4 vPosition;
uniform mat4 mvp;
void main()
{
    gl_Position = mvp * vPosition;
}
```

❑   `fshader`: This is the fragment shader source:

```
void main()
{
    gl_FragColor = vec4(0.0,0.5,0.0,1.0);
}
```

6. Add the `Android.mk` file under the `jni` folder as follows. Note that we must link to OpenGL ES 2.0 by `LOCAL_LDLIBS := -1GLESv2`:

```
LOCAL_PATH := $(call my-dir)
include $(CLEAR_VARS)
LOCAL_MODULE     := CubeG2
LOCAL_SRC_FILES := matrix.cpp Cube.cpp CubeG2.cpp
LOCAL_LDLIBS := -1GLESv2 -llog
include $(BUILD_SHARED_LIBRARY)
```

7. Add the following line before `<application>...</application>` in the `AndroidManifest.xml` file, which indicates that the Android application uses the OpenGL ES 2.0 feature:

```
<uses-feature android:glEsVersion="0x00020000"
android:required="true" />
```

8. Build the Android NDK application and run it on an Android device. The app will display a cube and we can touch to rotate the cube:

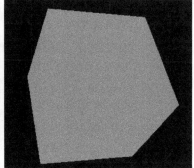

## How it works...

The sample project renders a 3D cube using OpenGL ES 2.0. OpenGL ES 2.0 provides a programmable pipeline, where a vertex shader and fragment shader can be supplied to control how the vertex and fragment are processed:

▶   **Vertex shader**: It's executed for every vertex. Transforms, lighting, texture mapping, and so on are usually performed using it.

> ▸ **Fragment shader**: It's executed for every fragment produced by the rasterizer. A typical processing is to adding colors to every fragment.

Shaders are programmed using OpenGL Shading Language, which is discussed next.

## OpenGL Shading Language (GLSL)

Here, we briefly introduce GLSL.

> ▸ **Data types**: They are of four main types, including `bool`, `int`, `float`, and `sampler`. There are also vector types for the first three types—`bvec2`, `bvec3`, `bvec4` refer to 2D, 3D, and 4D boolean vectors. `ivec2`, `ivec3`, and `ivec4` represent integer vectors. `vec2`, `vec3`, and `vec4` refer to floating point vectors. **Samplers** are used for texture sampling and have to be uniform.

> ▸ **Attributes, uniforms, and varyings**: A shader includes three types of inputs and outputs, including uniforms, attributes, and varyings. All three types have to be global:

> > ❑ **Uniform**: It is of read-only type and doesn't need to be changed during rendering. For example, light position.

> > ❑ **Attribute**: It is of read-only type and is only available as an input to the vertex shader. It changes for every vertex. For example, vertex position.

> > ❑ **Varying**: It is used to pass data from the vertex shader to the fragment shader. It is readable and writable in the vertex shader, but only readable in the fragment shader.

> ▸ **Built-in types**: GLSL has various built-in attributes, uniforms, and varyings for shaders. We highlight a few of them as follows:

> > ❑ `gl_Vertex`: It is an attribute—a 4D vector representing the vertex position.

> > ❑ `gl_Color`: It is an attribute—a 4D vector representing the vertex color.

> > ❑ `gl_ModelViewMatrix`: It is an uniform—the 4x4 model view matrix.

> > ❑ `gl_ModelViewProjectionMatrix`: It is a uniform. The 4x4 model view projection matrix.

> > ❑ `gl_Position`: It is only available as vertex shader output. It's a 4D vector representing the final processed vertex position.

> > ❑ `gl_FragColor`: It is only available as fragment shader output. It's a 4D vector representing the final color to be written to the frame buffer.

## How to use shader:

In our sample project, the vertex shader program simply multiplies every cube vertex with the model-view-projection matrix, and the fragment shader sets green color to every fragment. The following steps should be followed to use the shader source code:

1. **Create Shaders**: The following OpenGL ES 2.0 methods are called:

   - glCreateShader: It creates a GL_VERTEX_SHADER or GL_FRAGMENT_SHADER shader. A non-zero value is returned by it, by which the shader can be referenced.

   - glShaderSource: It puts the source code in a shader object. The source code stored previously will be completely replaced.

   - glCompileShader: It compiles the source code of the shader object.

2. **Create a program and attach the shaders**: The following methods are called:

   - glCreateProgram: It creates an empty program object to which shaders can be attached. Program objects essentially provide a mechanism to link everything needed to be executed together.

   - glAttachShader: It attaches a shader to a program object.

   - glLinkProgram: It links a program object. If any GL_VERTEX_SHADER objects are attached to the program object, they will be used to create an executable running on the vertex processor. If any GL_FRAGMENT_SHADER shaders are attached, they will be used to create an executable running on the fragment processor.

3. **Use the program**: We use the following calls to pass data to shaders and perform OpenGL operations:

   - glUseProgram: A program object as part of current rendering state is installed

   - glGetAttribLocation: It returns an attribute variable's location

   - glVertexAttribPointer: It specifies the location and data format of the array of generic vertex attributes to use at rendering

   - glEnableVertexAttribArray: It enables a vertex attribute array

   - glGetUniformLocation: It returns a uniform variable's location

   - glUniform: It specifies the value of a uniform variable

   - glDrawArrays: It renders primitives from the array data.

## There's more...

The sample project performs model-view transform and projection transform through **matrix operations**. The details of these transforms are tedious and not within the scope of this book, therefore we won't cover them here. However, detailed comments are provided along with the code. Interested readers could also easily find online resources about these operations.

# Displaying graphics with EGL

Besides the `GLSurfaceView` display mechanism we described in the previous recipe, it is also possible to display OpenGL graphics using EGL.

## Getting ready

Readers are recommended to read the *Drawing 3D Graphics and Lighting up the Scene with OpenGL ES 1.x API* recipe before going through this one.

## How to do it...

The following steps describe how to create an Android project that demonstrates the usage of EGL:

1.  Create an Android application named `EGLDemo`. Set the package name as `cookbook.chapter4.egl`. Please refer to the *Loading native libraries and registering native methods* recipe in *Chapter 2, Java Native Interface*, if you want more detailed instructions.

2.  Right-click on the project `EGLDemo`, select **Android Tools | Add Native Support**.

3.  Add two Java files, namely `EGLDemoActivity.java` and `MySurfaceView.java`. `EGLDemoActivity.java` sets `ContentView` as an instance of `MySurfaceView`, and starts and stops rendering at the Android activity callback functions:

```
… …
public void onCreate(Bundle savedInstanceState) {
super.onCreate(savedInstanceState);
myView = new MySurfaceView(this);
this.setContentView(myView);
}
protected void onResume() {
super.onResume();
myView.startRenderer();
}
… …
```

```
protected void onStop() {
super.onStop();
myView.destroyRender();
}
... ...
```

4. `MySurfaceView.java` performs role similar to `GLSurfaceView`. It interacts with the the native renderer to manage the display surface and handle touch events:

```
public class MySurfaceView extends SurfaceView implements
SurfaceHolder.Callback {
... ...
public MySurfaceView(Context context) {
super(context);
this.getHolder().addCallback(this);
}
... ...
public boolean onTouchEvent(final MotionEvent event) {
float x = event.getX();
float y = event.getY();
switch (event.getAction()) {
case MotionEvent.ACTION_MOVE:
    float dx = x - mPreviousX;
    float dy = y - mPreviousY;
    mAngleX += dx * TOUCH_SCALE_FACTOR;
    mAngleY += dy * TOUCH_SCALE_FACTOR;
    naRequestRenderer(mAngleX, mAngleY);
}
mPreviousX = x;
mPreviousY = y;
return true;
}
@Override
public void surfaceChanged(SurfaceHolder holder, int format, int
width,int height) {
naSurfaceChanged(holder.getSurface());
}
@Override
public void surfaceCreated(SurfaceHolder holder) {}
@Override
public void surfaceDestroyed(SurfaceHolder holder) {
naSurfaceDestroyed();
}
}
```

5. The following code should be added to the `jni` folder:

- ❑ **Cube.cpp and Cube.h**: Use the OpenGL 1.x API to draw a 3D cube.

- ❑ **OldRenderMethods.cpp and OldRenderMethods.h**: Initialize OpenGL 1.x, perform transforms, draw graphics, and so on. This is similar to the corresponding methods in *the Drawing 3D Graphics in OpenGL 1.x recipe*.

- ❑ **Renderer.cpp and Renderer.h**: Simulate `android.opengl. GLSurfaceView.Renderer`. It sets up the EGL context, manages the display, and so on.

- ❑ **renderAFrame**: It sets the event type, and then signals the rendering thread to handle the event:

```
void Renderer::renderAFrame(float pAngleX, float pAngleY) {
pthread_mutex_lock(&mMutex);
mAngleX = pAngleX; mAngleY = pAngleY;
mRendererEvent = RTE_DRAW_FRAME;
pthread_mutex_unlock(&mMutex);
pthread_cond_signal(&mCondVar);
}
```

- ❑ **renderThreadRun**: It runs in a separate thread to handle various events, including surface change, draw a frame, and so on:

```
void Renderer::renderThreadRun() {
    bool ifRendering = true;
    while (ifRendering) {
        pthread_mutex_lock(&mMutex);
        pthread_cond_wait(&mCondVar, &mMutex);
        switch (mRendererEvent) {
        ... ...
            case RTE_DRAW_FRAME:
                mRendererEvent = RTE_NONE;
                pthread_mutex_unlock(&mMutex);
                if (EGL_NO_DISPLAY!=mDisplay) {
            naDrawGraphics(mAngleX, mAngleY);
            eglSwapBuffers(mDisplay, mSurface);
            }
                }
                break;
            ......
        }
    }
}
```

❏ `initDisplay`: It sets up the EGL context:

```cpp
bool Renderer::initDisplay() {
const EGLint attribs[] = {
    EGL_SURFACE_TYPE, EGL_WINDOW_BIT,
    EGL_BLUE_SIZE, 8,
    EGL_GREEN_SIZE, 8,
    EGL_RED_SIZE, 8,
    EGL_NONE};
EGLint width, height, format;
EGLint numConfigs;
EGLConfig config;
EGLSurface surface;
EGLContext context;
EGLDisplay display = eglGetDisplay(EGL_DEFAULT_DISPLAY);
eglInitialize(display, 0, 0);
eglChooseConfig(display, attribs, &config, 1, &numConfigs);
eglGetConfigAttrib(display, config, EGL_NATIVE_VISUAL_ID,
&format);
ANativeWindow_setBuffersGeometry(mWindow, 0, 0, format);
surface = eglCreateWindowSurface(display, config, mWindow,
NULL);
context = eglCreateContext(display, config, NULL, NULL);
if (eglMakeCurrent(display, surface, surface, context) ==
EGL_FALSE) {
    return -1;
}
eglQuerySurface(display, surface, EGL_WIDTH, &width);
eglQuerySurface(display, surface, EGL_HEIGHT, &height);
  … ...
}
```

❏ `EGLDemo.cpp`: It registers the native methods and wraps the native code. The following two methods are used:

`naSurfaceChanged`: It gets the native window associated with a Java Surface object and initializes EGL and OpenGL:

```cpp
void naSurfaceChanged(JNIEnv* env, jclass clazz, jobject
pSurface) {
gWindow = ANativeWindow_fromSurface(env, pSurface);
gRenderer->initEGLAndOpenGL1x(gWindow);
}
```

naRequestRenderer: It renders a frame, which is called by the `touch` event handler in `MySurfaceView`:

```
void naRequestRenderer(JNIEnv* env, jclass clazz, float
pAngleX, float pAngleY) {
gRenderer->renderAFrame(pAngleX, pAngleY);
}
```

6. Add the `Android.mk` file under the `jni` folder with the following content:

```
LOCAL_PATH := $(call my-dir)
include $(CLEAR_VARS)
LOCAL_MODULE := EGLDemo
LOCAL_SRC_FILES := Cube.cpp OldRenderMethods.cpp Renderer.cpp
EGLDemo.cpp
LOCAL_LDLIBS := -llog -landroid -lEGL -lGLESv1_CM
include $(BUILD_SHARED_LIBRARY)
```

7. Build the Android NDK application and run it on an Android device. The app will display a cube, which we can touch to rotate it:

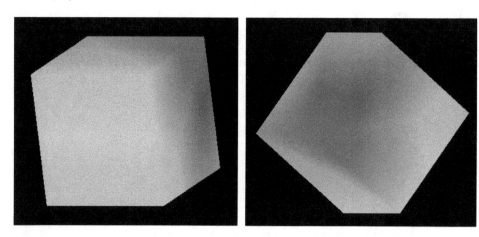

## How it works...

EGL is an interface between OpenGL ES and the underlying native window system. According to Khronos EGL web page (http://www.khronos.org/egl), graphics context management, surface binding, and rendering synchronization for rendering with other Khronos 2D and 3D APIs, including OpenGL ES are handled by it.

**EGL** is a cross-platform API widely used in embedded systems, including Android and iPhone (the EGL implementation from Apple is called **EAGL**). Many desktop platforms also support EGL. Different implementations may not be 100 percent compatible, but the porting effort will usually not be substantial for the EGL code.

The following steps describe how to set up and manipulate EGL and its integration with OpenGL:

1. **Get and initialize the display connection**: EGL needs to know where the content should be displayed, therefore we will need to get a display connection and initialize it. This is done using the following two methods:

   - `eglGetDisplay`: It obtains the EGL display connection for the native display. If the input argument is `EGL_DEFAULT_DISPLAY`, a default display connection is returned.

   - `eglInitialize`: It initializes an EGL display connection obtained by `eglGetDisplay`.

2. **Configure EGL**: This is done through `eglChooseConfig`.

   `eglChooseConfig` returns a list of EGL frame buffer configurations that match the requirements specified by the `attrib_list` argument. The attribute is an array with pairs of attributes and corresponding desired values, and it is terminated by `EGL_NONE`. In our code, we simply specify `EGL_SURFACE_TYPE` as `EGL_WINDOW_BIT`, and color components sizes as 8 bit.

3. **Create a render surface where the display content will be placed**: This is done through `eglCreateWindowSurface`.

   `eglCreateWindowSurface`, given the EGL display connection, the EGL frame buffer configuration and native window returns a new EGL window surface.

   In our code, we start from `SurfaceView` and pass its associated `android.view.Surface` value to the native code. In the native code, we obtain its native window, and finally create the EGL window surface for OpenGL drawing.

4. **Create the EGL rendering context and make it the current**: This is done by `eglCreateContext` and `eglMakeCurrent`.

   - `eglCreateContext`: It creates a new EGL rendering context, which is used to render into the EGL draw surface.

   - `eglMakeCurrent`: It attaches an EGL context to the EGL draw and read surfaces. In our code, the created window surface is used as both the read and draw surface.

5. **OpenGL drawing**: This is covered in previous recipes.

6. **Swap the EGL surface internal buffers to display the content**: This is done by the `eglSwapBuffers` call.

   `eglSwapBuffers` posts the EGL surface color buffer to a native window. This effectively displays the drawing content on the screen.

   EGL internally maintains two buffers. The content of the front buffer is displayed, while the drawing can be done on the back buffer. At the time we decided to display the new drawing, we swap the two buffers.

7. At time we want to stop rendering. Release the EGL context, destroy the EGL surface, and terminate the EGL display connection:

   ❑ `eglMakeCurrent` with `EGL_NO_SURFACE` and `EGL_NO_CONTEXT` releases the current context

   ❑ `eglDestroySurface` destroys an EGL surface

   ❑ `eglTerminate` terminates the EGL display connection

## Window management

Our code uses the Android native window management API calls to obtain a native window and configure it. The following methods are called:

▶ `ANativeWindow_fromSurface`: It returns a native window associated with the Java surface object. The returned reference should be passed to `ANativeWindow_release` to ensure there's no leaking.

▶ `ANativeWindow_setBuffersGeometry`: It sets the size and format of window buffers. In our code, we specified width and height as `0`, in which case the window's base value will be used.

Note that we'll need to link to the Android library in the `Android.mk` file (`LOCAL_LDLIBS := -landroid`), because it is a part of the Android native application API, which we will cover more in the next chapter.

## There's more...

The renderer runs an event loop in a separate thread. We used the **POSIX thread** (`pthreads`) calls to create a native thread, synchronize it with the main thread, and so on. We'll cover `pthread` in detail in *Chapter 6, Android NDK Multithreading*.

# 5

# Android Native Application API

In this chapter we will cover the following recipes:

- ▸ Creating a native activity with the native_activity.h interface
- ▸ Creating a native activity with the Android native app glue
- ▸ Managing native windows at Android NDK
- ▸ Detecting and handling input events at Android NDK
- ▸ Accessing sensors at Android NDK
- ▸ Managing assets at Android NDK

## Introduction

Thanks to the Android native application APIs, it is possible to write an Android application with pure native code since Android API level 9 (Android 2.3, Gingerbread). That is, not a single line of Java code is needed. The Android native APIs are defined in several header files under the `<NDK root>/platforms/android-<API level>/arch-arm/usr/include/android/` folder. Based on the features provided by the functions defined in these header files, they can be grouped as follows:

- ▸ Activity lifecycle management:
  - ❑ `native_activity.h`
  - ❑ `looper.h`

- ▶ Windows management:
  - ❑ `rect.h`
  - ❑ `window.h`
  - ❑ `native_window.h`
  - ❑ `native_window_jni.h`

- ▶ Input (including key and motion events) and sensor events:
  - ❑ `input.h`
  - ❑ `keycodes.h`
  - ❑ `sensor.h`

- ▶ Assets, configuration, and storage management:
  - ❑ `configuration.h`
  - ❑ `asset_manager.h`
  - ❑ `asset_manager_jni.h`
  - ❑ `storage_manager.h`
  - ❑ `obb.h`

In addition, Android NDK also provides a static library named **native app glue** to help create and manage native activities. The source code of this library can be found under the `sources/android/native_app_glue/` directory.

In this chapter, we will first introduce the creation of a native activity with the simple callback model provided by `native_acitivity.h`, and the more complicated but flexible two-threaded model enabled by the native app glue library. We will then discuss window management at Android NDK, where we will draw something on the screen from the native code. Input events handling and sensor accessing are introduced next. Lastly, we will introduce asset management, which manages the files under the `assets` folder of our project. Note that the APIs covered in this chapter can be used to get rid of the Java code completely, but we don't have to do so. The *Managing assets at Android NDK* recipe provides an example of using the asset management API in a mixed-code Android project.

Before we start, it is important to keep in mind that although no Java code is needed in a native activity, the Android application still runs on Dalvik VM, and a lot of Android platform features are accessed through JNI. The Android native application API just hides the Java world for us.

# Creating a native activity with the native_activity.h interface

The Android native application API allows us to create a native activity, which makes writing Android apps in pure native code possible. This recipe introduces how to write a simple Android application with pure C/C++ code.

## Getting ready

Readers are expected to have basic understanding of how to invoke JNI functions. *Chapter 2, Java Native Interface*, covers JNI in detail and readers are recommended to read the chapter or at least the following recipes before going through the current one:

- *Manipulating strings in Android NDK*
- *Calling instance and static methods in NDK*

## How to do it...

The following steps to create a simple Android NDK application without a single line of Java code:

1. Create an Android application named `NativeActivityOne`. Set the package name as `cookbook.chapter5.nativeactivityone`. Please refer to the *Loading native libraries and registering native methods* recipe of *Chapter 2, Java Native Interface*, if you want more detailed instructions.

2. Right-click on the `NativeActivityOne` project, select **Android Tools | Add Native Support**.

3. Change the `AndroidManifest.xml` file as follows:

```
<manifest xmlns:android="http://schemas.android.com/apk/res/
android"
    package="cookbook.chapter5.nativeactivityone"
    android:versionCode="1"
    android:versionName="1.0">
    <uses-sdk android:minSdkVersion="9"/>
    <application android:label="@string/app_name"
        android:icon="@drawable/ic_launcher"
        android:hasCode="true">
<activity android:name="android.app.NativeActivity"
        android:label="@string/app_name"
        android:configChanges="orientation|keyboardHidden">
            <meta-data android:name="android.app.lib_name"
                android:value="NativeActivityOne" />
```

```
        <intent-filter>
          <action android:name="android.intent.action.MAIN" />
          <category android:name="android.intent.category.
LAUNCHER" />
        </intent-filter>
      </activity>
    </application>

</manifest>
```

We should ensure that the following are set correctly in the preceding file:

- ❑ The activity name must be set to `android.app.NativeActivity`.

- ❑ The value of the `android.app.lib_name` metadata must be set to the native module name without the `lib` prefix and `.so` suffix.

- ❑ `android:hasCode` needs to be set to `true`, which indicates that the application contains code. Note that the documentation in `<NDK root>/docs/NATIVE-ACTIVITY.HTML` gives an example of the `AndroidManifest.xml` file with `android:hasCode` set to `false`, which will not allow the application to start.

4. Add two files named `NativeActivityOne.cpp` and `mylog.h` under the `jni` folder. The `ANativeActivity_onCreate` method should be implemented in `NativeActivityOne.cpp`. The following is an example of the implementation:

```
void ANativeActivity_onCreate(ANativeActivity* activity,
        void* savedState, size_t savedStateSize) {
  printInfo(activity);
  activity->callbacks->onStart = onStart;
  activity->callbacks->onResume = onResume;
  activity->callbacks->onSaveInstanceState = onSaveInstanceState;
  activity->callbacks->onPause = onPause;
  activity->callbacks->onStop = onStop;
  activity->callbacks->onDestroy = onDestroy;
  activity->callbacks->onWindowFocusChanged =
onWindowFocusChanged;
  activity->callbacks->onNativeWindowCreated =
onNativeWindowCreated;
  activity->callbacks->onNativeWindowResized =
onNativeWindowResized;
  activity->callbacks->onNativeWindowRedrawNeeded =
onNativeWindowRedrawNeeded;
  activity->callbacks->onNativeWindowDestroyed =
onNativeWindowDestroyed;
  activity->callbacks->onInputQueueCreated = onInputQueueCreated;
  activity->callbacks->onInputQueueDestroyed =
onInputQueueDestroyed;
```

```
    activity->callbacks->onContentRectChanged =
onContentRectChanged;
    activity->callbacks->onConfigurationChanged =
onConfigurationChanged;
    activity->callbacks->onLowMemory = onLowMemory;
    activity->instance = NULL;
}
```

5.  Add the `Android.mk` file under the `jni` folder:

```
LOCAL_PATH := $(call my-dir)
include $(CLEAR_VARS)
LOCAL_MODULE     := NativeActivityOne
LOCAL_SRC_FILES := NativeActivityOne.cpp
LOCAL_LDLIBS     := -landroid -llog
include $(BUILD_SHARED_LIBRARY)
```

6.  Build the Android application and run it on an emulator or a device. Start a terminal and display the logcat output using the following:

```
$ adb logcat -v time NativeActivityOne:I *:S
```

Alternatively, you can use the logcat view at Eclipse to see the logcat output.

When the application starts, you should be able to see the following logcat output:

```
09-20 00:36:54.842 I/NativeActivityOne( 596): internalDataPath: /data/data/cookbook.chapter5.nativeactivityone/files
09-20 00:36:54.852 I/NativeActivityOne( 596): externalDataPath: /mnt/sdcard/Android/data/cookbook.chapter5.nativeactivityone/files
09-20 00:36:54.852 I/NativeActivityOne( 596): Activity toString: android.app.NativeActivity@412aa230
09-20 00:36:54.852 I/NativeActivityOne( 596): SDK version: 15
09-20 00:36:54.852 I/NativeActivityOne( 596): -----ANativeActivity_onCreate
09-20 00:36:54.862 I/NativeActivityOne( 596): -----onStart: 0x19f4f8
09-20 00:36:54.862 I/NativeActivityOne( 596): -----onResume: 0x19f4f8
09-20 00:36:54.922 I/NativeActivityOne( 596): -----onInputQueueCreated: 0x19f4f8 -- 0x1adf30
09-20 00:36:55.002 I/NativeActivityOne( 596): ----onNativeWindowCreated: 0x19f4f8 -- 0x1cc070
09-20 00:36:55.002 I/NativeActivityOne( 596): -----onNativeWindowResized: 0x19f4f8 -- 0x1cc070
09-20 00:36:55.012 I/NativeActivityOne( 596): -----onContentRectChanged: 0x19f4f8 -- 0xbead2640
09-20 00:36:55.032 I/NativeActivityOne( 596): -----onNativeWindowRedrawNeeded: 0x19f4f8 -- 0x1cc070
```

As shown in the screenshot, a few Android activity lifecycle callback functions are executed. We can manipulate the phone to cause other callbacks being executed. For example, long pressing the home button and then pressing the back button will cause the `onWindowFocusChanged` callback to be executed.

## How it works...

In our example, we created a simple, "pure" native application to output logs when the Android framework calls into the callback functions defined by us. The "pure" native application is not really pure native. Although we did not write a single line of Java code, the Android framework still runs some Java code on Dalvik VM.

Android framework provides an `android.app.NativeActivity.java` class to help us create a "native" activity. In a typical Java activity, we extend `android.app.Activity` and overwrite the activity lifecycle methods. `NativeActivity` is also a subclass of `android.app.Activity` and does similar things. At the start of a native activity, `NativeActivity.java` will call `ANativeActivity_onCreate`, which is declared in `native_activity.h` and implemented by us. In the `ANativeActivity_onCreate` method, we can register our callback methods to handle activity lifecycle events and user inputs. At runtime, `NativeActivity` will invoke these native callback methods when the corresponding events occurred.

In a word, `NativeActivity` is a wrapper that hides the managed Android Java world for our native code, and exposes the native interfaces defined in `native_activity.h`.

**The ANativeActivity data structure**: Every callback method in the native code accepts an instance of the `ANativeActivity` structure. Android NDK defines the `ANativeActivity` data structure in `native_acitivity.h` as follows:

```
typedef struct ANativeActivity {
    struct ANativeActivityCallbacks* callbacks;
    JavaVM* vm;
    JNIEnv* env;
    jobject clazz;
    const char* internalDataPath;
    const char* externalDataPath;
    int32_t sdkVersion;
    void* instance;
    AAssetManager* assetManager;
} ANativeActivity;
```

The various attributes of the preceding code are explained as follows:

- `callbacks`: It is a data structure that defines all the callbacks that the Android framework will invoke with the main UI thread.

- `vm`: It is the application process' global Java VM handle. It is used in some JNI functions.

- `env`: It is a `JNIEnv` interface pointer. `JNIEnv` is used through local storage data (refer to the *Manipulating strings in Android NDK* recipe in *Chapter 2, Java Native Interface*, for more details), so this field is only accessible through the main UI thread.

- `clazz`: It is a reference to the `android.app.NativeActivity` object created by the Android framework. It can be used to access fields and methods in the `android.app.NativeActivity` Java class. In our code, we accessed the `toString` method of `android.app.NativeActivity`.

- `internalDataPath`: It is the internal data directory path for the application.

- `externalDataPath`: It is the external data directory path for the application.

internalDataPath and externalDataPath are NULL at Android 2.3.x. This is a known bug and has been fixed since Android 3.0. If we are targeting devices lower than Android 3.0, then we need to find other ways to get the internal and external data directories.

- ▸ sdkVersion: It is the Android platform's SDK version code. Note that this refers to the version of the device/emulator that runs the app, not the SDK version used in our development.

- ▸ instance: It is not used by the framework. We can use it to store user-defined data and pass it around.

- ▸ assetManager: It is the a pointer to the app's instance of the asset manager. We will need it to access assets data. We will discuss it in more detail in the *Managing assets at Android NDK* recipe of this chapter.

## There's more...

The native_activity.h interface provides a simple single thread callback mechanism, which allows us to write an activity without Java code. However, this single thread approach infers that we must quickly return from our native callback methods. Otherwise, the application will become unresponsive to user actions (for example, when we touch the screen or press the **Menu** button, the app does not respond because the GUI thread is busy executing the callback function).

A way to solve this issue is to use multiple threads. For example, many games take a few seconds to load. We will need to offload the loading to a background thread, so that the UI can display the loading progress and be responsive to user inputs. Android NDK comes with a static library named android_native_app_glue to help us in handling such cases. The details of this library are covered in the *Creating a native activity with the Android native app glue* recipe.

A similar problem exists at Java activity. For example, if we write a Java activity that searches the entire device for pictures at onCreate, the application will become unresponsive. We can use AsyncTask to search and load pictures in the background, and let the main UI thread display a progress bar and respond to user inputs.

# Creating a native activity with the Android native app glue

The previous recipe described how the interface defined in `native_activity.h` allows us to create native activity. However, all the callbacks defined are invoked with the main UI thread, which means we cannot do heavy processing in the callbacks.

Android SDK provides `AsyncTask`, `Handler`, `Runnable`, `Thread`, and so on, to help us handle things in the background and communicate with the main UI thread. Android NDK provides a static library named `android_native_app_glue` to help us execute callback functions and handle user inputs in a separate thread. This recipe will discuss the `android_native_app_glue` library in detail.

## Getting ready

The `android_native_app_glue` library is built on top of the `native_activity.h` interface. Therefore, readers are recommended to read the *Creating a native activity with the native_activity.h interface* recipe before going through this one.

## How to do it...

The following steps create a simple Android NDK application based on the `android_native_app_glue` library:

1.  Create an Android application named `NativeActivityTwo`. Set the package name as `cookbook.chapter5.nativeactivitytwo`. Please refer to the *Loading native libraries and registering native methods* recipe of *Chapter 2, Java Native Interface*, if you want more detailed instructions.

2.  Right-click on the `NativeActivityTwo` project, select **Android Tools | Add Native Support**.

3.  Change the `AndroidManifest.xml` file as follows:

```
<manifest xmlns:android="http://schemas.android.com/apk/res/
android"
    package="cookbook.chapter5.nativeactivitytwo"
    android:versionCode="1"
    android:versionName="1.0">
    <uses-sdk android:minSdkVersion="9"/>
    <application android:label="@string/app_name"
        android:icon="@drawable/ic_launcher"
        android:hasCode="true">
    <activity android:name="android.app.NativeActivity"
        android:label="@string/app_name"
```

```
            android:configChanges="orientation|keyboardHidden">
                <meta-data android:name="android.app.lib_name"
                   android:value="NativeActivityTwo" />
                <intent-filter>
                    <action android:name="android.intent.action.MAIN" />
                    <category android:name="android.intent.category.
LAUNCHER" />
                </intent-filter>
            </activity>
        </application>
    </manifest>
```

4. Add two files named `NativeActivityTwo.cpp` and `mylog.h` under the `jni` folder. `NativeActivityTwo.cpp` is shown as follows:

```
#include <jni.h>
#include <android_native_app_glue.h>
#include "mylog.h"
void handle_activity_lifecycle_events(struct android_app* app,
int32_t cmd) {
   LOGI(2, "%d: dummy data %d", cmd, *((int*)(app->userData)));
}
void android_main(struct android_app* app) {
   app_dummy();     // Make sure glue isn't stripped.
   int dummyData = 111;
   app->userData = &dummyData;
   app->onAppCmd = handle_activity_lifecycle_events;
   while (1) {
      int ident, events;
      struct android_poll_source* source;
if ((ident=ALooper_pollAll(-1, NULL, &events, (void**)&source)) >=
0) {
        source->process(app, source);
      }
   }
}
```

5. Add the `Android.mk` file under the `jni` folder:

```
LOCAL_PATH := $(call my-dir)
include $(CLEAR_VARS)
LOCAL_MODULE     := NativeActivityTwo
LOCAL_SRC_FILES := NativeActivityTwo.cpp
LOCAL_LDLIBS     := -llog -landroid
LOCAL_STATIC_LIBRARIES := android_native_app_glue
include $(BUILD_SHARED_LIBRARY)
$(call import-module,android/native_app_glue)
```

6. Build the Android application and run it on an emulator or device. Start a terminal and display the logcat output by using the following command:

```
adb logcat -v time NativeActivityTwo:I *:S
```

When the application starts, you should be able to see the following logcat output and the device screen will shows a black screen:

```
09-21 23:29:29.122 I/NativeActivityTwo(31875): 10: dummy data 111
09-21 23:29:29.122 I/NativeActivityTwo(31875): 11: dummy data 111
09-21 23:29:29.122 I/NativeActivityTwo(31875): 0: dummy data 111
09-21 23:29:29.152 I/NativeActivityTwo(31875): 1: dummy data 111
09-21 23:29:29.162 I/NativeActivityTwo(31875): 6: dummy data 111
```

On pressing the back button, the following output will be shown:

```
09-21 23:29:32.966 I/NativeActivityTwo(31875): 13: dummy data 111
09-21 23:29:33.016 I/NativeActivityTwo(31875): 7: dummy data 111
09-21 23:29:33.046 I/NativeActivityTwo(31875): 2: dummy data 111
09-21 23:29:33.206 I/NativeActivityTwo(31875): 14: dummy data 111
09-21 23:29:33.206 I/NativeActivityTwo(31875): 0: dummy data 111
09-21 23:29:33.206 I/NativeActivityTwo(31875): 15: dummy data 111
```

## How it works...

This recipe demonstrates how the `android_native_app_glue` library is used to create a native activity.

The following steps should be followed to use the `android_native_app_glue` library:

▶ Implement a function named `android_main`. This function should implement an event loop, which will poll for events continuously. This method will run in the background thread created by the library.

▶ Two event queues are attached to the background thread by default, including the activity lifecycle event queue and the input event queue. When polling events using the looper created by the library, you can identify where the event is coming from, by checking the returned identifier (either `LOOPER_ID_MAIN` or `LOOPER_ID_INPUT`). It is also possible to attach additional event queues to the background thread.

▶ When an event is returned, the data pointer will point to an `android_poll_source` data structure. We can call the process function of this structure. The process is a function pointer, which points to `android_app->onAppCmd` for activity lifecycle events, and `android_app->onInputEvent` for input events. We can provide our own processing functions and direct the corresponding function pointers to these functions.

In our example, we implement a simple function named `handle_activity_lifecycle_events` and point the `android_app->onAppCmd` function pointer to it. This function simply prints the `cmd` value and the user data passed along with the `android_app` data structure. `cmd` is defined in `android_native_app_glue.h` as an enum. For example, when the app starts, the `cmd` values are `10`, `11`, `0`, `1`, and `6`, which correspond to `APP_CMD_START`, `APP_CMD_RESUME`, `APP_CMD_INPUT_CHANGED`, `APP_CMD_INIT_WINDOW`, and `APP_CMD_GAINED_FOCUS` respectively.

**android_native_app_glue Library Internals**: The source code of the `android_native_app_glue` library can be found under the `sources/android/native_app_glue` folder of Android NDK. It only consists of two files, namely `android_native_app_glue.c` and `android_native_app_glue.h`. Let's first describe the flow of the code and then discuss some important aspects in detail.

 Since the source code for `native_app_glue` is provided, we can modify it if necessary, although in most cases it won't be necessary.

`android_native_app_glue` is built on top of the `native_activity.h` interface. As shown in the following code (extracted from `sources/android/native_app_glue/android_native_app_glue.c`). It implements the `ANativeActivity_onCreate` function, where it registers the callback functions and calls the `android_app_create` function. Note that the returned `android_app` instance is pointed by the `instance` field of the native activity, which can be passed to various callback functions:

```
void ANativeActivity_onCreate(ANativeActivity* activity,
        void* savedState, size_t savedStateSize) {
    LOGV("Creating: %p\n", activity);
    activity->callbacks->onDestroy = onDestroy;
    activity->callbacks->onStart = onStart;
    activity->callbacks->onResume = onResume;

    ... ...
    activity->callbacks->onNativeWindowCreated =
onNativeWindowCreated;
    activity->callbacks->onNativeWindowDestroyed =
onNativeWindowDestroyed;
    activity->callbacks->onInputQueueCreated = onInputQueueCreated;
    activity->callbacks->onInputQueueDestroyed =
onInputQueueDestroyed;
    activity->instance = android_app_create(activity, savedState,
savedStateSize);
}
```

The android_app_create function (shown in the following code snippet) initializes an instance of the android_app data structure, which is defined in android_native_app_ glue.h. This function creates a unidirectional pipe for inter-thread communication. After that, it spawns a new thread (let's call it **background thread** thereafter) to run the android_ app_entry function with the initialized android_app data as the input argument. The main thread will wait for the background thread to start and then return:

```
static struct android_app* android_app_create(ANativeActivity*
activity, void* savedState, size_t savedStateSize) {
    struct android_app* android_app = (struct android_app*)
malloc(sizeof(struct android_app));
    memset(android_app, 0, sizeof(struct android_app));
    android_app->activity = activity;

    pthread_mutex_init(&android_app->mutex, NULL);
    pthread_cond_init(&android_app->cond, NULL);
    ......
    int msgpipe[2];
    if (pipe(msgpipe)) {
        LOGE("could not create pipe: %s", strerror(errno));
        return NULL;
    }
    android_app->msgread = msgpipe[0];
    android_app->msgwrite = msgpipe[1];

    pthread_attr_t attr;
    pthread_attr_init(&attr);
    pthread_attr_setdetachstate(&attr, PTHREAD_CREATE_DETACHED);
    pthread_create(&android_app->thread, &attr, android_app_entry,
android_app);
    // Wait for thread to start.
    pthread_mutex_lock(&android_app->mutex);
    while (!android_app->running) {
        pthread_cond_wait(&android_app->cond, &android_app->mutex);
    }
    pthread_mutex_unlock(&android_app->mutex);
    return android_app;
}
```

The background thread starts with the android_app_entry function (as shown in the following code snippet), where a looper is created. Two event queues will be attached to the looper. The activity lifecycle events queue is attached to the android_app_entry function. When the activity's input queue is created, the input queue is attached (to the android_ app_pre_exec_cmd function of android_native_app_glue.c). After attaching the activity lifecycle event queue, the background thread signals the main thread it is already running. It then calls a function named android_main with the android_app data. android_main is the function we need to implement, as shown in our sample code. It must run in a loop until the activity exits:

```
static void* android_app_entry(void* param) {
    struct android_app* android_app = (struct android_app*)param;
    ... ...
  //Attach life cycle event queue with identifier LOOPER_ID_MAIN
    android_app->cmdPollSource.id = LOOPER_ID_MAIN;
    android_app->cmdPollSource.app = android_app;
    android_app->cmdPollSource.process = process_cmd;
    android_app->inputPollSource.id = LOOPER_ID_INPUT;
    android_app->inputPollSource.app = android_app;
    android_app->inputPollSource.process = process_input;
    ALooper* looper = ALooper_prepare(ALOOPER_PREPARE_ALLOW_NON_
CALLBACKS);
    ALooper_addFd(looper, android_app->msgread, LOOPER_ID_MAIN,
ALOOPER_EVENT_INPUT, NULL, &android_app->cmdPollSource);
    android_app->looper = looper;

    pthread_mutex_lock(&android_app->mutex);
    android_app->running = 1;
    pthread_cond_broadcast(&android_app->cond);
    pthread_mutex_unlock(&android_app->mutex);
    android_main(android_app);
    android_app_destroy(android_app);
    return NULL;
}
```

The following diagram indicates how the main and background thread work together to create the multi-threaded native activity:

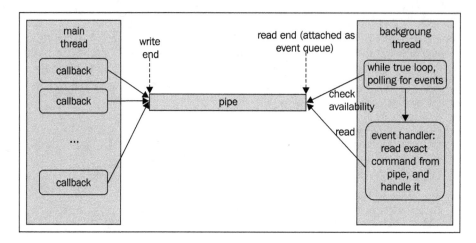

We use the activity lifecycle event queue as an example. The main thread invokes the callback functions, which simply writes to the write end of the pipe, while true loop implemented in the android_main function will poll for events. Once an event is detected, the function calls the event handler, which reads the exact command from the read end of the pipe and handles it. The android_native_app_glue library implements all the main thread stuff and part of the background thread stuff for us. We only need to supply the polling loop and the event handler as illustrated in our sample code.

**Pipe**: The main thread creates a unidirectional pipe in the android_app_create function by calling the pipe method. This method accepts an array of two integers. After the function is returned, the first integer will be set as the file descriptor referring to the read end of the pipe, while the second integer will be set as the file descriptor referring to the write end of the pipe.

A pipe is usually used for **Inter-process Communication (IPC)**, but here it is used for communication between the main UI thread and the background thread created at android_app_entry. When an activity lifecycle event occurs, the main thread will execute the corresponding callback function registered at ANativeActivity_onCreate. The callback function simply writes a command to the write end of the pipe and then waits for a signal from the background thread. The background thread is supposed to poll for events continuously and once it detects a lifecycle event, it will read the exact event from the read end of the pipe, signal the main thread to unblock and handle the events. Because the signal is sent right after receiving the command and before actual processing of the events, the main thread can return from the callback function quickly without worrying about the possible long processing of the events.

Different operating systems have different implementations for the pipe. The pipe implemented by Android system is "half-duplex", where communication is unidirectional. That is, one file descriptor can only write, and the other file descriptor can only read. Pipes in some operating system is "full-duplex", where the two file descriptors can both read and write.

**Looper** is an event tracking facility, which allows us to attach one or more event queues for an event loop of a thread. Each event queue has an associated file descriptor. An **event** is data available on a file descriptor. In order to use a looper, we need to include the `android/looper.h` header file.

The library attaches two event queues for the event loop to be created by us in the background thread, including the activity lifecycle event queue and the input event queue. The following steps should be performed in order to use a looper:

1.  **Create or obtain a looper associated with the current thread**: This is done by the `ALooper_prepare` function:

    ```
    ALooper* ALooper_prepare(int opts);
    ```

    This function prepares a looper associated with the calling thread and returns it. If the looper doesn't exist, it creates one, associates it with the thread, and returns it.

2.  **Attach an event queue**: This is done by `ALooper_addFd`. The function has the following prototype:

    ```
    int ALooper_addFd(ALooper* looper, int fd, int ident, int events,
    ALooper_callbackFunc callback, void* data);
    ```

    The function can be used in two ways. Firstly, if `callback` is set to `NULL`, the `ident` set will be returned by `ALooper_pollOnce` and `ALooper_pollAll`. Secondly, if `callback` is non-NULL, then the callback function will be executed and `ident` is ignored. The `android_native_app_glue` library uses the first approach to attach a new event queue to the looper. The input argument `fd` indicates the file descriptor associated with the event queue. `ident` is the identifier for the events from the event queue, which can be used to classify the event. The identifier must be bigger than zero when `callback` is set to `NULL`. `callback` is set to `NULL` in the library source code, and `data` points to the private data that will be returned along with the identifier at polling.

    In the library, this function is called to attach the activity lifecycle event queue to the background thread. The input event queue is attached using the input queue specific function `AInputQueue_attachLooper`, which we will discuss in the *Detecting and handling input events at NDK* recipe.

3. **Poll for events**: This can be done by either one of the following two functions:

```
int ALooper_pollOnce(int timeoutMillis, int* outFd, int*
outEvents, void** outData);
int ALooper_pollAll(int timeoutMillis, int* outFd, int* outEvents,
void** outData);
```

These two methods are equivalent when `callback` is set to NULL in `ALooper_addFd`. They have the same input arguments. `timeoutMillis` specifies the timeout for polling. If it is set to zero, then the functions return immediately; if it is set to negative, they will wait indefinitely until an event occurs. The functions return the identifier (greater than zero) when an event occurs from any input queues attached to the looper. In this case, `outFd`, `outEvents`, and `outData` will be set to the file descriptor, poll events, and data associated with the event. Otherwise, they will be set to NULL.

4. **Detach event queues**: This is done by the following function:

```
int ALooper_removeFd(ALooper* looper, int fd);
```

It accepts the looper and file descriptor associated with the event queue, and detaches the queue from the looper.

# Managing native windows at Android NDK

The previous recipes in this chapter provided simple examples with the logcat output only. This recipe will discuss how to manage the native window at Android NDK.

## Getting ready

Readers are recommended to read the following recipes before going through this one:

▶ *Creating a native activity with the native_activity.h interface*

▶ *Creating a native activity with the Android native app glue*

Also recall that native window management has been introduced briefly in the *Displaying graphics with EGL* recipe in *Chapter 4, Android NDK OpenGL ES API*.

## How to do it...

The following steps create the sample application:

1. Create an Android application named `NativeWindowManagement`. Set the package name as `cookbook.chapter5.nativewindowmanagement`. Please refer to the *Loading native libraries and registering native methods* recipe of *Chapter 2, Java Native Interface*, if you want more detailed instructions.

2. Right-click on the `NativeWindowManagement` project, select **Android Tools | Add Native Support**.

3. Update `AndroidManifest.xml`. Please refer to previous recipe or the downloaded code for details. Note that the metadata `android.app.lib_name` must have its value as `NativeWindowManagement`.

4. Add two files named `NativeWindowManagement.cpp` and `mylog.h` under the `jni` folder. `NativeWindowManagement.cpp` is modified based on previous recipe. The following code snippet shows the updated part:

```cpp
void drawSomething(struct android_app* app) {
  ANativeWindow_Buffer lWindowBuffer;
  ANativeWindow* lWindow = app->window;
  ANativeWindow_setBuffersGeometry(lWindow, 0, 0, WINDOW_FORMAT_
RGBA_8888);
  if (ANativeWindow_lock(lWindow, &lWindowBuffer, NULL) < 0) {
    return;
  }
  memset(lWindowBuffer.bits, 0, lWindowBuffer.
stride*lWindowBuffer.height*sizeof(uint32_t));
  int sqh = 150, sqw = 100;
  int wst = lWindowBuffer.stride/2 - sqw/2;
  int wed = wst + sqw;
  int hst = lWindowBuffer.height/2 - sqh/2;
  int hed = hst + sqh;
  for (int i = hst; i < hed; ++i) {
    for (int j = wst; j < wed; ++j) {
      ((char*)(lWindowBuffer.bits))[(i*lWindowBuffer.stride +
j)*sizeof(uint32_t)] = (char)255;        //R
      ((char*)(lWindowBuffer.bits))[(i*lWindowBuffer.stride +
j)*sizeof(uint32_t) + 1] = (char)0;      //G
      ((char*)(lWindowBuffer.bits))[(i*lWindowBuffer.stride +
j)*sizeof(uint32_t) + 2] = (char)0;      //B
      ((char*)(lWindowBuffer.bits))[(i*lWindowBuffer.stride +
j)*sizeof(uint32_t) + 3] = (char)255;    //A
    }
  }
  ANativeWindow_unlockAndPost(lWindow);
}

void handle_activity_lifecycle_events(struct android_app* app,
int32_t cmd) {
  LOGI(2, "%d: dummy data %d", cmd, *((int*)(app->userData)));
  switch (cmd) {
  case APP_CMD_INIT_WINDOW:
    drawSomething(app);
    break;
  }
}
```

5. Add the `Android.mk` file under the `jni` folder, which is similar to the one used in the previous recipe. You just need to replace the module name as `NativeWindowManagement` and the source file as `NativeWindowManagement.cpp`.

6. Build the Android application and run it on an emulator or device. Start a terminal and display the logcat output by using the following command:

```
$ adb logcat -v time NativeWindowManagement:I *:S
```

When the application starts, we will see the following logcat:

```
09-22 11:35:00.953 I/NativeWindowManagement( 3443): 10: dummy data 111
09-22 11:35:00.963 I/NativeWindowManagement( 3443): 11: dummy data 111
09-22 11:35:00.963 I/NativeWindowManagement( 3443): 0: dummy data 111
09-22 11:35:00.983 I/NativeWindowManagement( 3443): 1: dummy data 111
09-22 11:35:00.983 I/NativeWindowManagement( 3443): window initialized
09-22 11:35:00.983 I/NativeWindowManagement( 3443): height 800, width 480, stride 480
09-22 11:35:00.993 I/NativeWindowManagement( 3443): drawing square width: 190:290, height: 325:475
09-22 11:35:01.013 I/NativeWindowManagement( 3443): 6: dummy data 111
```

The device screen will display a red rectangle at the center of the screen, as follows:

## How it works...

The NDK interface for native window management is defined in the `window.h`, `rect.h`, `native_window_jni.h`, and `native_window.h` header files. The first two simply define some constants and data structures. `native_window_jni.h` defines a single function named `ANativeWindow_fromSurface`, which helps us to obtain a native window from a Java surface object. We have illustrated this function in the *Displaying graphics with EGL* recipe in *Chapter 4, Android NDK OpenGL ES API*. Here, we focus on the functions provided in `native_window.h`.

Perform the following steps to draw a square on the phone screen:

1. **Set the window buffer format and size:**This is done by the `ANativeWindow_ setBuffersGeometry` function:

   ```
   int32_t ANativeWindow_setBuffersGeometry(ANativeWindow* window,
   int32_t width, int32_t height, int32_t format);
   ```

   This function updates the native window buffer associated with the native window referred by the window input argument. The window size and format are changed according to the rest of the input arguments. Three formats are defined in `native_window.h`, including `WINDOW_FORMAT_RGBA_8888`, `WINDOW_FORMAT_ RGBX_8888`, and `WINDOW_FORMAT_RGB_565`. If the size or the format is set to `0`, then the native window's base value will be used.

2. **Lock the window's next drawing surface**: This is done by the `ANativeWindow_ lock` function:

   ```
   int32_t ANativeWindow_lock(ANativeWindow* window, ANativeWindow_
   Buffer* outBuffer,  ARect* inOutDirtyBounds);
   ```

   After this call is returned, the input argument `outBuffer` will refer to the window buffer for drawing.

3. **Clear the buffer**: This is optional. Sometimes we may just want to overwrite a part of the window buffer. In our example, we called `memset` to set all the data to `0`.

4. **Draw something to the buffer**: In our example, we first calculate the start and end width and height of the rectangle, and then set the red and alpha bytes of the rectangle area as `255`. This will show us a red rectangle.

5. **Unlock the window's drawing surface and post the new buffer to display**: This is done with the `ANativeWindow_unlockAndPost` function:

   ```
   int32_t ANativeWindow_unlockAndPost(ANativeWindow* window);
   ```

# Detecting and handling input events at Android NDK

Input events are essential for user interaction in Android apps. This recipe discusses how to detect and handle input events in Android NDK.

## Getting ready

We will further develop the example in last recipe. Please read the *Managing native windows at Android NDK* recipe before going through this one.

## How to do it...

The following steps create a sample application, which detects and handles input events at the native code:

1. Create an Android application named `NativeInputs`. Set the package name as `cookbook.chapter5.nativeinputs`. Please refer to the *Loading native libraries and registering native methods* recipe of *Chapter 2, Java Native Interface*, if you want more detailed instructions.

2. Right-click on the `NativeInputs` project, select **Android Tools | Add Native Support.**

3. Update `AndroidManifest.xml`. Please refer to previous recipe or the downloaded code for details. Note that the metadata `android.app.lib_name` must have a value as `NativeInputs`.

4. Add two files named `NativeInputs.cpp` and `mylog.h` under the `jni` folder. `NativeInputs.cpp` is modified based on the previous recipe. Let us see a part of its code here:

   ❑ `handle_input_events`: This is the event handler method for input events. Note that when a motion event with move action (`AINPUT_EVENT_TYPE_MOTION`) is detected, we update `app->userData` and set `app->redrawNeeded` to 1:

```
int mPreviousX = -1;
int32_t handle_input_events(struct android_app* app,
AInputEvent* event) {
   int etype = AInputEvent_getType(event);
   switch (etype) {
   case AINPUT_EVENT_TYPE_KEY:
... ...
      break;
   case AINPUT_EVENT_TYPE_MOTION:
      int32_t action, posX, pointer_index;
```

```
    action = AMotionEvent_getAction(event);
    pointer_index = (action&AMOTION_EVENT_ACTION_POINTER_
INDEX_MASK) >> AMOTION_EVENT_ACTION_POINTER_INDEX_SHIFT;
    posX = AMotionEvent_getX(event, pointer_index);
    if (action == AMOTION_EVENT_ACTION_MOVE) {
      int xMove = posX - mPreviousX;
      USERDATA* userData = (USERDATA*)app->userData;
      userData->xMove = xMove;
      app->redrawNeeded = 1;
    }
    mPreviousX = posX;
    break;
  }
}
```

❑ `android_main`: We update the while true loop. When `app->redrawNeeded` is set, we redraw the rectangle:

```
void android_main(struct android_app* app) {
… ...
while (1) {
    int ident, events;
    struct android_poll_source* source;
    if ((ident=ALooper_pollOnce(app->redrawNeeded?0:-1,
NULL, &events, (void**)&source)) >= 0) {
        if (NULL!=source) {
          source->process(app, source);
        }
        if (app->redrawNeeded) {
          drawSomething(app);
        }
    }
  }
}
```

5. Add the `Android.mk` file under the `jni` folder, which is similar to previous recipe. We just need to replace the module name as `NativeInputs` and the source file as `NativeInputs.cpp`.

6. Build the Android application and run it on an emulator or device. We can move a figure across the screen to see the rectangle moving horizontally:

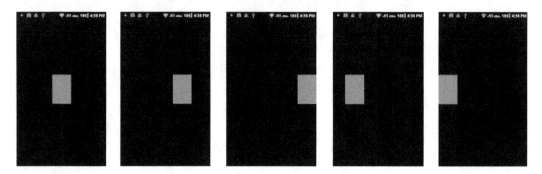

## How it works...

This recipe discusses input events handling with the `android_native_app_glue` library at Android NDK.

**Input event queue in android_native_app_glue**: `android_native_app_glue` attaches the input event queue for us by default.

1. When the input queue is created for an activity, the `onInputQueueCreated` callback is called on the main thread, which writes `APP_CMD_INPUT_CHANGED` to the write end of the pipe we described in previous recipe. The background thread will receive the command and call `AInputQueue_attachLooper` the function to attach the input queue to the background thread looper.

2. When an input event occurs, it will be handled by `process_input` (the function pointer `source->process` in the while true loop we called points to `process_input` if the event is an input event). Inside `process_input`, `AInputQueue_getEvent` is firstly called to retrieve the event. Then, `AInputQueue_preDispatchEvent` is called to send the key for pre-dispatching. This could possibly result in it being consumed by the current **Input Method Editor** (**IME**) before the app. Followed by this is the `android_app->onInputEvent`, which is a function pointer-pointing to an event handler provided by us. If no event handler is provided by us, it's set to `NULL`. After that, `AInputQueue_finishEvent` is called to indicate that event handling is over.

3. Lastly, when the input queue is destroyed, the `onInputQueueDestroyed` callback is called on the main thread, which also writes `APP_CMD_INPUT_CHANGED`. The background thread will read the command and call a function named `AInputQueue_detachLooper` to detach the input queue from the thread looper.

**Event handler**: In the `handle_input_events` function, we first called `AInputEvent_getType` to get the input event type. The `android/input.h` header file defines two input event types, namely `AINPUT_EVENT_TYPE_KEY` and `AINPUT_EVENT_TYPE_MOTION`. The first event type indicates that the input event is a key event, while the second one indicates that it is a motion event.

We called `AKeyEvent_getAction`, `AKeyEvent_getFlags`, and `AKeyEvent_getKeyCode` to get the action, flags, and key code of a key event and printed a string to describe it. On the other hand, we called `AMotionEvent_getAction` and `AMotionEvent_getX` to get the action and the x position of a motion event. Note that the `AMotionEvent_getX` function requires the second input argument as the pointer index. The pointer index is obtained by using the following code:

```
pointer_index = (action&AMOTION_EVENT_ACTION_POINTER_INDEX_MASK) >>
AMOTION_EVENT_ACTION_POINTER_INDEX_SHIFT;
```

There are a lot more input event functions, which can be found at `andoid/input.h`.

# Accessing sensors at Android NDK

Many Android devices have built-in sensors to detect and measure motion, orientation, and other environmental conditions. It is possible to access sensors in Android NDK. This recipe will discuss how to do it in detail.

## Getting ready

The example provided in this recipe is based on the sample code in the previous two recipes. Readers are recommended to read them first:

- *Managing native windows at Android NDK*
- *Detecting and handling input events at Android NDK*

## How to do it...

The following steps develop the sample Android application, which demonstrates how to access sensors from Android NDK:

1. Create an Android application named `nativesensors`. Set the package name as `cookbook.chapter5.nativesensors`. Please refer to the *Loading native libraries and registering native methods* recipe of *Chapter 2, Java Native Interface*, if you want more detailed instructions.

2. Right-click on the `nativesensors` project, select **Android Tools | Add Native Support**.

3. Update `AndroidManifest.xml`. Please refer to previous recipe or the downloaded code for details. Note that the metadata `android.app.lib_name` must have a value as `nativesensors`.

4. Add two files named `nativesensors.cpp` and `mylog.h` under the `jni` folder. Let's show a part of the code in `nativesensors.cpp`.

   ❑ `handle_activity_lifecycle_events`: This function handles activity lifecycle events. We enable the sensor when the activity is in focus and disable it when the activity loses its focus. This saves the battery life by avoiding reading sensors when our activity is not in focus:

```cpp
void handle_activity_lifecycle_events(struct android_app*
app, int32_t cmd) {
  USERDATA* userData;
  switch (cmd) {
….. . .
    case APP_CMD_SAVE_STATE:
      // save current state
      userData = (USERDATA*)(app->userData);
      app->savedState = malloc(sizeof(SAVED_USERDATA));
      *((SAVED_USERDATA*)app->savedState) = userData-
>drawingData;
      app->savedStateSize = sizeof(SAVED_USERDATA);
      break;
    case APP_CMD_GAINED_FOCUS:
      userData = (USERDATA*)(app->userData);
      if (NULL != userData->accelerometerSensor) {
        ASensorEventQueue_enableSensor(userData-
>sensorEventQueue,
            userData->accelerometerSensor);
        ASensorEventQueue_setEventRate(userData-
>sensorEventQueue,
            userData->accelerometerSensor, (1000L/60)*1000);
      }
      break;
    case APP_CMD_LOST_FOCUS:
      USERDATA userData = *(USERDATA*) app->userData;
      if (NULL!=userData.accelerometerSensor) {
ASensorEventQueue_disableSensor(userData.sensorEventQueue,
userData.accelerometerSensor);
      }
      break;
  }
}
```

❑ `android_main`: We continuously poll for events and handle the sensor
events identified by the `LOOPER_ID_USER` identifier:

```
void android_main(struct android_app* app) {
… ...
while (0==app->destroyRequested) {
  int ident, events;
  struct android_poll_source* source;
  if ((ident=ALooper_pollOnce(-1, NULL, &events,
(void**)&source)) >= 0) {
      if (LOOPER_ID_USER == ident) {
      ASensorEvent event;
      while (ASensorEventQueue_getEvents(userData.
sensorEventQueue,
            &event, 1) > 0) {
        int64_t currentTime = get_time();
        … ...
        if ((currentTime - lastTime) > TIME_THRESHOLD) {
          long diff = currentTime - lastTime;
          float speedX = (event.acceleration.x - lastX)/
diff*10000;
          float speedY = (event.acceleration.y - lastY)/
diff*10000;
          float speedZ = (event.acceleration.z - lastZ)/
diff*10000;
          float speed = fabs(speedX + speedY + speedZ);
… ...
        }
      }
      }
    }
  }
```

```
ASensorManager_destroyEventQueue(userData.sensorManager, userData.
sensorEventQueue);
}
```

5. Add the `Android.mk` file under the `jni` folder, which is similar to the one used in the
previous recipe. We just need to replace the module name as `nativesensors` and
the source file as `nativesensors.cpp`.

6.  Build the Android application and run it on an emulator or device. We can shake the device to see the rectangle moving horizontally:

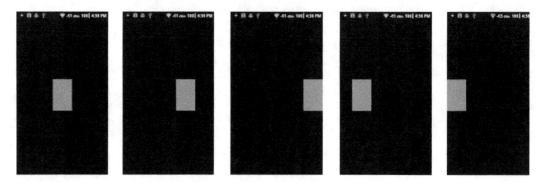

## How it works...

In our example, we used the accelerometer sensor to detect phone shaking. Then, based on the phone shaking speed, we move the red rectangle to one side of the phone screen. Once the rectangle reaches an edge of the phone screen, it starts to move to the other edge.

The example code provides a simple algorithm to determine whether a shake has happened or not. More complex and accurate algorithms exist and can be implemented. We can also adjust the SHAKE_TIMEOUT and SHAKE_COUNT_THRESHOLD constants to fine tune the algorithm.

The important part of the example is how to access sensors. Let's summarize the steps:

1.  **Get a reference to the sensor manager**: This is done by using the following function:

    ```
    ASensorManager* ASensorManager_getInstance();
    ```

2.  **Get the default sensor of a given type**: We can also get a list of all available sensors. This is done by using the following two functions respectively:

    ```
    ASensor const* ASensorManager_getDefaultSensor(ASensorManager*
    manager, int type);
    int ASensorManager_getSensorList(ASensorManager* manager,
    ASensorList* list);
    ```

    The available types are defined in android/sensor.h. In our example, we print all sensor names and types but only use ASENSOR_TYPE_ACCELEROMETER.

3.  **Create a new sensor queue and attach it to the looper of the thread**: This is done by using the ASensorManager_createEventQueue function as follows:

    ```
    ASensorEventQueue* ASensorManager_createEventQueue(ASensorMana
    ger* manager, ALooper* looper, int ident, ALooper_callbackFunc
    callback, void* data);
    ```

The usage of this function is similar to the usagw of the `ALooper_addFd` function in the *Creating a native activity with the Android native app glue* recipe and `AInputQueue_attachLooper` in the *Detecting and handling input events at Android NDK* recipe. In our example, we set the `ident` as `LOOPER_ID_USER`. Note that we may also define a new looper ID by changing the code of `android_native_app_glue.h` and setting it here.

4. **Enable and configure the sensor**:

```
int ASensorEventQueue_enableSensor(ASensorEventQueue* queue,
ASensor const* sensor);
int ASensorEventQueue_setEventRate(ASensorEventQueue* queue,
ASensor const* sensor, int32_t usec);
```

The first function enables the sensor referred by the sensor input argument. The second function sets the delivery rate of the events, in microseconds, for the sensor referred by the sensor input argument. In our example, we called these two functions when the activity gained focus.

5. **Poll for events and get the available events from the queue**: The polling is done by calling `ALooper_pollOnce`, as shown in the previous recipe. If the event identifier returned is `LOOPER_ID_USER`, we know that it is a sensor event and we can use the following function to get it:

```
ssize_t ASensorEventQueue_getEvents(ASensorEventQueue* queue,
ASensorEvent* events, size_t count);
```

`count` indicates the maximum number of available events we want to get. In our example, we set it to `1`. It is also possible to define an array of `ASensorEvent` and get multiple events at one time.

6. **Handle sensor events**: The sensor event is represented by the `ASensorEvent` data structure, which can be found at `android/sensor.h` (the exact path to the file is `<Android NDK root dir>/platforms/android-<version>/arch-arm/ usr/include/android/sensor.h`). In our example, we accessed the acceleration readings at the x, y, and z axes, and used the readings to determine if a phone shake has happened.

7. **Disable the sensor**: After you are done accessing the sensors, you can disable it with the following function:

```
int ASensorEventQueue_disableSensor(ASensorEventQueue* queue,
ASensor const* sensor);
```

8. **Destroy the sensor event queue and free all resources associated with it**:

```
int ASensorManager_destroyEventQueue(ASensorManager* manager,
ASensorEventQueue* queue);
```

# Managing assets at Android NDK

Assets provide a way for Android apps to include various types of files, including text, image, audio, video, and so on. This recipe discusses how to load asset files from Android NDK.

## Getting ready

We will modify the example we developed in the *Mapping texture in OpenGL ES 1.x* recipe in *Chapter 4, Android NDK OpenGL ES API*. Readers are suggested to read through the recipe or take a look at the code first.

## How to do it...

The following steps describe how the sample application is developed:

1. Create an Android application named `NativeAssets`. Set the package name as `cookbook.chapter5.nativeassets`. Please refer to the *Loading native libraries and registering native methods* recipe of *Chapter 2, Java Native Interface*, if you want more detailed instructions.

2. Right-click on the `NativeAssets` project, select **Android Tools | Add Native Support**.

3. Add three Java files, namely `MyActivity.java`, `MySurfaceView.java`, and `MyRenderer.java` under the `cookbook.chapter5.nativeassets` package. The first two files are identical to the corresponding files in the *Mapping texture in OpenGL ES 1.x* recipe in *Chapter 4, Android NDK OpenGL ES API*. The last file is slightly changed, where the `naLoadTexture` native method signature is updated as follows:

   ```
   private static native void naLoadTexture(AssetManager
   pAssetManager);
   ```

   In the `onSurfaceCreated` method, we called the native method by passing a Java `AssetManager` instance:

   ```
   naLoadTexture(mContext.getAssets());
   ```

4. Create two folders under the `jni` folder, namely `dice` and `libpng-1.5.12`. In the `libpng-1.5.12` folder, we place the source files of `libpng`, which can be downloaded from `http://sourceforge.net/projects/libpng/files/`.

In the `dice` folder, we add the `Cube.cpp`, `Cube.h`, `mylog.h`, and `DiceG1.cpp` files. The first three files are the same as the example in the *Mapping texture in OpenGL ES 1.x* recipe in *Chapter 4, Android NDK OpenGL ES API*. The `DiceG1. cpp` file is updated by adding procedures to read `.png` assets files from the `assets` folder. Let's show a part of the updated code:

- ❑ `readPng`: It is the callback function used at `png_set_read_fn`. It reads the data from the `asset` file:

```
void readPng(png_structp pPngPtr, png_bytep pBuf, png_size_t
pCount) {
  AAsset* assetF = (AAsset*)png_get_io_ptr(pPngPtr);
  AAsset_read(assetF, pBuf, pCount);
}
```

- ❑ `naLoadTexture`: It reads all the `.png` files under the `assets` top-level directory and loads the data to OpenGL for texture mapping:

```
void naLoadTexture(JNIEnv* env, jclass clazz, jobject
pAssetManager) {
  AAssetManager* assetManager = AAssetManager_fromJava(env,
pAssetManager);
  AAssetDir* texDir = AAssetManager_openDir(assetManager,
"");
  const char* texFn;
  int pId = 0;
  while (NULL != (texFn = AAssetDir_
getNextFileName(texDir))) {
    AAsset* assetF = AAssetManager_open(assetManager, texFn,
AASSET_MODE_UNKNOWN);
    //read the png header
    png_byte header[8];
    png_byte *imageData;
    ......
    if (8 != AAsset_read(assetF, header, 8)) {
      goto FEND;
    }
    ......
    //init png reading by setting a read callback
    png_set_read_fn(pngPtr, assetF, readPng);
    ......
    // Loads image data into OpenGL.
    glTexImage2D(GL_TEXTURE_2D, 0, format, width, height, 0,
format, type, imageData);
```

```
FEND:
        AAsset_close(assetF);
        pId++;
    }
  AAssetDir_close(texDir);
}
```

5. Add an `Android.mk` file under `jni`, `jni/dice`, and `jni/libpng-1.5.12` respectively. The `Android.mk` file under the `jni` top-level folder is as follows. This simply instructs the Android build system to include the `Android.mk` files under each sub-directory under the `jni` folder:

```
LOCAL_PATH := $(call my-dir)
include $(call all-subdir-makefiles)
```

The `Android.mk` file under the `jni/libpng-1.5.12` folder is as follows. This compiles `libpng` as a local static library:

```
LOCAL_PATH := $(call my-dir)
include $(CLEAR_VARS)
LOCAL_CFLAGS :=
LOCAL_MODULE     := libpng
LOCAL_SRC_FILES :=\
  png.c \
  pngerror.c \
  pngget.c \
  pngmem.c \
  pngpread.c \
  pngread.c \
  pngrio.c \
  pngrtran.c \
  pngrutil.c \
  pngset.c \
  pngtrans.c \
  pngwio.c \
  pngwrite.c \
  pngwtran.c \
  pngwutil.c
LOCAL_LDLIBS := -lz
include $(BUILD_STATIC_LIBRARY)
```

The `Android.mk` file under the `jni/dice` folder is as follows:

```
LOCAL_PATH := $(call my-dir)
include $(CLEAR_VARS)
LOCAL_MODULE     := DiceG1NativeAssets
LOCAL_C_INCLUDES := $(LOCAL_PATH)/../libpng-1.5.12/
```

```
LOCAL_STATIC_LIBRARIES := libpng
LOCAL_SRC_FILES := Cube.cpp DiceG1.cpp
LOCAL_LDLIBS := -lGLESv1_CM -llog -landroid -lz
include $(BUILD_SHARED_LIBRARY)
```

6. Build the Android NDK application and run it on an Android device. The app will display a cube textured as a dice; this is the same as what we have seen in *Chapter 4, Android NDK OpenGL ES API*.

## How it works...

In the example, we load the `.png` files from the `assets` folder and used them as OpenGL textures. You can use the following steps to read `assets`:

1. **Get a native AAssetManager object from the Java AssetManager object**: This is done by the `AAssetManager_fromJava` function, which is defined in `asset_manager_jni.h`.

2. **Open an asset directory**: This is done by AAssetManager_openDir.

   ```
   AAssetDir* AAssetManager_openDir(AAssetManager* mgr, const char* dirName);
   ```

   To open the top-level directory "assets", we set dirName to "". For the subdirectories, we will need to supply the directory name.

3. **Get an asset file name**:

   ```
   const char* AAssetDir_getNextFileName(AAssetDir* assetDir);
   ```

   Iterate over the files under the `asset` directory referred by the input argument `assetDir`. If all files have been returned or there are no files, NULL is returned.

4. **Open an asset file**: This is done by using AAssetManager_open:

   ```
   AAsset* AAssetManager_open(AAssetManager* mgr, const char* filename, int mode);
   ```

The filename should be set to the `asset` file name, where `mode` can be one of the following:

- ❑ AASSET_MODE_UNKNOWN: Not known how the data is to be accessed
- ❑ AASSET_MODE_RANDOM: Read chunks, and seek forward and backward
- ❑ AASSET_MODE_STREAMING: Read sequentially, with an occasional forward seek
- ❑ AASSET_MODE_BUFFER: Attempt to load contents into memory, for fast small reads

5. **Read the asset file**: This is done by using `AAsset_read`.

```
int AAsset_read(AAsset* asset, void* buf, size_t count);
```

The input argument `buf` refers to the location where the data is placed after reading, and `count` indicates the number of bytes we want to read. The actual number of bytes read is returned and may differ from `count`.

6. **Close the asset file**: This is done by using the `AAsset_close` function.

7. **Close the asset directory**: This is done by using the `AAssetDir_close` function.

## There's more...

In this example, we built `libpng` as a local static library. This is necessary to read the `.png` files, because Android NDK does not provide APIs to access `.png` files. We will discuss how to develop Android NDK applications with existing libraries in *Chapter 8, Porting and Using the Existing*.

# 6
# Android NDK Multithreading

In this chapter we will cover:

- ▶ Creating and terminating native threads at Android NDK
- ▶ Synchronizing native threads with mutex at Android NDK
- ▶ Synchronizing native threads with conditional variables at Android NDK
- ▶ Synchronizing native threads with reader/writer locks at Android NDK
- ▶ Synchronizing native threads with semaphore at Android NDK
- ▶ Scheduling native threads at Android NDK
- ▶ Managing data for native threads at Android NDK

## Introduction

Most non-trivial Android apps use more than one thread, therefore multithreaded programming is essential to Android development. At Android NDK, **POSIX Threads (pthreads)** is bundled in Android's Bionic C library to support multithreading. This chapter mainly discusses the API functions defined in the `pthread.h` and `semaphore.h` header files, which can be found under the `platforms/android-<API level>/arch-arm/usr/include/` folder of Android NDK.

We will first introduce thread creation and termination. Synchronization is important in all multithreaded applications, therefore we discuss four commonly used synchronization techniques at Android NDK with four recipes, including mutex, conditional variables, reader/writer locks, and semaphore. We then illustrate thread scheduling and finally describe how to manage data for threads.

Being a practical book, we will not cover the theories behind multithreaded programming. Readers are expected to understand the basics of multithreading, including concurrency, mutual exclusion, deadlock, and so on.

In addition, pthreads programming is a complicated topic and there are books written solely for pthreads programming. This chapter will only focus on the essentials at the context of Android NDK programming. Interested readers can refer to the book *Pthreads Programming: A POSIX Standard for Better Multiprocessing*, by *Bradford Nicols, Dick Buttlar*, and *Jacqueline Proulx Farrell* for more information.

# Creating and terminating native threads at Android NDK

This recipe discusses how to create and terminate native threads at Android NDK.

## Getting ready...

Readers are expected to know how to create an Android NDK project. We can refer to the *Writing a Hello NDK program* recipe in *Chapter 1, Hello NDK*, for detailed instructions.

## How to do it...

The following steps describe how to create a simple Android application with multiple native threads:

1. Create an Android application named `NativeThreadsCreation`. Set the package name as `cookbook.chapter6.nativethreadscreation`. Refer to the *Loading native libraries and registering native methods* recipe in *Chapter 2, Java Native Interface* for more detailed instructions.

2. Right-click on the project **NativeThreadsCreation**, select **Android Tools | Add Native Support**.

3. Add a Java file named `MainActivity.java` under package `cookbook.chapter6.nativethreadscreation`. This Java file simply loads the native library `NativeThreadsCreation` and calls the native `jni_start_threads` method.

4. Add `mylog.h` and `NativeThreadsCreation.cpp` files under the `jni` folder. The `mylog.h` file contains the Android native `logcat` utility functions, while the `NativeThreadsCreation.cpp` file contains the native code to start multiple threads. A part of the code is shown next.

The `jni_start_threads` function starts two threads and waits for the two threads to terminate:

```
void jni_start_threads() {
  pthread_t th1, th2;
  int threadNum1 = 1, threadNum2 = 2;
  int ret;
  ret = pthread_create(&th1, NULL, run_by_thread,
(void*)&threadNum1);
  ret = pthread_create(&th2, NULL, run_by_thread,
(void*)&threadNum2);
  void *status;
  ret = pthread_join(th1, &status);
  int* st = (int*)status;
  LOGI(1, "thread 1 end %d %d", ret, *st);
  ret = pthread_join(th2, &status);
  st = (int*)status;
  LOGI(1, "thread 2 end %d %d", ret, *st);
}
```

The `run_by_thread` function is executed to the native threads:

```
int retStatus;
void *run_by_thread(void *arg) {
  int cnt = 3, i;
  int* threadNum = (int*)arg;
  for (i = 0; i < cnt; ++i) {
    sleep(1);
    LOGI(1, "thread %d: %d", *threadNum, i);
  }
  if (1 == *threadNum) {
    retStatus = 100;
    return (void*)&retStatus;
  } else if (2 == *threadNum) {
    retStatus = 200;
    pthread_exit((void*)&retStatus);
  }
}
```

5. Add an `Android.mk` file in the `jni` folder with the following code:

```
LOCAL_PATH := $(call my-dir)
include $(CLEAR_VARS)
LOCAL_MODULE := NativeThreadsCreation
LOCAL_SRC_FILES := NativeThreadsCreation.cpp
LOCAL_LDLIBS := -llog
include $(BUILD_SHARED_LIBRARY)
```

6. Build and run the Android project, and use the following command to monitor the `logcat` output:

```
$ adb logcat -v time NativeThreadsCreation:I *:S
```

The following is a screenshot of the `logcat` output:

```
10-07 16:59:02.664 I/NativeThreadsCreation( 5599): thread 1 started
10-07 16:59:02.664 I/NativeThreadsCreation( 5599): thread 2 started
10-07 16:59:03.665 I/NativeThreadsCreation( 5599): thread 1: 0
10-07 16:59:03.665 I/NativeThreadsCreation( 5599): thread 2: 0
10-07 16:59:04.666 I/NativeThreadsCreation( 5599): thread 1: 1
10-07 16:59:04.666 I/NativeThreadsCreation( 5599): thread 2: 1
10-07 16:59:05.657 I/NativeThreadsCreation( 5599): thread 1: 2
10-07 16:59:05.657 I/NativeThreadsCreation( 5599): thread 1 end 0 100
10-07 16:59:05.657 I/NativeThreadsCreation( 5599): thread 2: 2
10-07 16:59:05.657 I/NativeThreadsCreation( 5599): thread 2 end 0 200
```

## How it works...

This recipe shows how to create and terminate threads at Android NDK.

### Build with pthreads

Traditionally, pthread is implemented as an external library and must be linked by providing a linker flag `-lpthread`. Android's Bionic C library has its own pthread implementation bundled in. Therefore, we do not use `-lpthread` in the `Android.mk` file in our project.

### Thread creation

As demonstrated in our code, a thread can be created with the `pthread_create` function, which has the following prototype:

```
int pthread_create(pthread_t *thread, const pthread_attr_t *attr, void
*(*start_routine)(void*), void *arg);
```

This function creates and starts a new thread with attributes specified by the `attr` input argument. If `attr` is set to `NULL`, default attributes are used. The `start_routine` argument points to the function to be executed by the newly created thread with `arg` as the input argument to the function. When the function returns, the `thread` input argument will point to a location where the thread ID is stored and the return value will be zero to indicate success, or other values to indicate error.

In our sample code, we created two threads to execute the `run_by_thread` function. We pass a pointer to an integer as input argument to the `run_by_thread` function.

## Thread termination

The thread is terminated after it returns from the `start_routine` function or we explicitly call `pthread_exit`. The `pthread_exit` function has the following prototype:

```
void pthread_exit(void *value_ptr);
```

This function terminates the calling thread and returns the value pointed by `value_ptr` to any successful `join` with the calling thread. This is also demonstrated in our sample code. We called `pthread_join` on both threads we created. The `pthread_join` function has the following prototype:

```
int pthread_join(pthread_t thread, void **value_ptr);
```

The function suspends the execution of the calling thread until the thread specified by the first input argument terminates. When the function returns successfully, the second argument can be used to retrieve the exit status of the terminated thread as demonstrated in our sample code.

In addition, the `logcat` screenshot that we have seen previously shows that calling return from a thread is equivalent to calling `pthread_exit`. Therefore, we can get the exit status when either method is called.

> `pthread_cancel` is not supported by Android Bionic C library pthread. Therefore, if we are porting code which uses `pthread_cancel`, we will need to refactor the code to get rid of it.

# Synchronizing native threads with mutex at Android NDK

This recipe discusses how to use pthread mutex at Android NDK.

## How to do it...

The following steps help to create an Android project that demonstrates the usage of pthread mutex:

1. Create an Android application named `NativeThreadsMutex`. Set the package name as `cookbook.chapter6.nativethreadsmutex`. Refer to the *Loading native libraries and registering native methods* recipe in *Chapter 2, Java Native Interface* for more detailed instructions.

2. Right-click on the project **NativeThreadsMutex**, select **Android Tools | Add Native Support**.

3. Add a Java file named `MainActivity.java` under the `cookbook.chapter6.nativethreadsmutex` package. This Java file simply loads the native `NativeThreadsMutex` library and calls the native `jni_start_threads` method.

4. Add two files named `mylog.h` and `NativeThreadsMutex.cpp` in the `jni` folder. `NativeThreadsMutex.cpp` contains the code to start two threads. The two threads will update a shared counter. A part of the code is shown as follows:

The `run_by_thread1` function is executed by the first native thread:

```
int cnt = 0;
int THR = 10;
void *run_by_thread1(void *arg) {
    int* threadNum = (int*)arg;
    while (cnt < THR) {
        pthread_mutex_lock(&mux1);
        while ( pthread_mutex_trylock(&mux2) ) {
            pthread_mutex_unlock(&mux1);  //avoid deadlock
            usleep(50000);  //if failed to get mux2, release mux1 first
            pthread_mutex_lock(&mux1);
        }
        ++cnt;
        LOGI(1, "thread %d: cnt = %d", *threadNum, cnt);
        pthread_mutex_unlock(&mux1);
        pthread_mutex_unlock(&mux2);
        sleep(1);
    }
}
```

The `run_by_thread2` function is executed by the second native thread:

```
void *run_by_thread2(void *arg) {
    int* threadNum = (int*)arg;
    while (cnt < THR) {
        pthread_mutex_lock(&mux2);
        while ( pthread_mutex_trylock(&mux1) ) {
            pthread_mutex_unlock(&mux2);  //avoid deadlock
            usleep(50000);   //if failed to get mux2, release mux1 first
            pthread_mutex_lock(&mux2);
        }
        ++cnt;
        LOGI(1, "thread %d: cnt = %d", *threadNum, cnt);
        pthread_mutex_unlock(&mux2);
        pthread_mutex_unlock(&mux1);
        sleep(1);
    }
}
```

5. Add an `Android.mk` file in the `jni` folder with the following content:

```
LOCAL_PATH := $(call my-dir)
include $(CLEAR_VARS)
LOCAL_MODULE := NativeThreadsMutex
LOCAL_SRC_FILES := NativeThreadsMutex.cpp
LOCAL_LDLIBS := -llog
include $(BUILD_SHARED_LIBRARY)
```

6. Build and run the Android project, and use the following command to monitor the `logcat` output.

```
$ adb logcat -v time NativeThreadsMutex:I *:S
```

The `logcat` output is shown as follows:

```
10-07 21:19:40.354 I/NativeThreadsMutex( 7931): thread 1 started
10-07 21:19:40.354 I/NativeThreadsMutex( 7931): thread 2 started
10-07 21:19:40.354 I/NativeThreadsMutex( 7931): thread 1: cnt = 1
10-07 21:19:40.354 I/NativeThreadsMutex( 7931): thread 2: cnt = 2
10-07 21:19:41.355 I/NativeThreadsMutex( 7931): thread 1: cnt = 3
10-07 21:19:41.355 I/NativeThreadsMutex( 7931): thread 2: cnt = 4
10-07 21:19:42.356 I/NativeThreadsMutex( 7931): thread 1: cnt = 5
10-07 21:19:42.356 I/NativeThreadsMutex( 7931): thread 2: cnt = 6
10-07 21:19:43.367 I/NativeThreadsMutex( 7931): thread 1: cnt = 7
10-07 21:19:43.367 I/NativeThreadsMutex( 7931): thread 2: cnt = 8
10-07 21:19:44.368 I/NativeThreadsMutex( 7931): thread 1: cnt = 9
10-07 21:19:44.368 I/NativeThreadsMutex( 7931): thread 2: cnt = 10
10-07 21:19:45.359 I/NativeThreadsMutex( 7931): thread 1 end 0
10-07 21:19:45.359 I/NativeThreadsMutex( 7931): thread 2 end 0
```

7. We also implemented a native method `jni_start_threads_dead` in `NativeThreadsMutex.cpp`, which can probably cause a deadlock (we may need to run the code a few times to produce the deadlock situation). If we call `jni_start_threads_dead` in `MainActivity.java`, the two threads will start and then block as shown in the following `logcat` output:

```
10-07 21:28:23.445 I/NativeThreadsMutex( 8322): thread 1 started
10-07 21:28:23.445 I/NativeThreadsMutex( 8322): thread 2 started
```

As indicated in this screenshot, the two threads cannot proceed after **started**.

## How it works...

The sample project demonstrates how to use mutex to synchronize native threads. We describe the details as follows:

### Initialize and destroy mutex

A mutex can be initialized with the `pthread_mutex_init` function, which has the following prototype:

```
int pthread_mutex_init(pthread_mutex_t *mutex, const pthread_
mutexattr_t *attr);
```

The input argument mutex is a pointer to the mutex to be initialized and `attr` indicates the attributes of mutex. If `attr` is set to `NULL`, the default attributes will be used. The function will return a zero if the mutex is initialized successfully and a non-zero value otherwise.

>  A macro `PTHREAD_MUTEX_INITIALIZER` is also defined in `pthread.h` to initialize a mutex with default attributes.

When we are done with the mutex, we can destroy it with the `pthread_mutex_destroy` function, which has the following prototype:

```
int pthread_mutex_destroy(pthread_mutex_t *mutex);
```

The input argument is a pointer pointing to the mutex to be destroyed.

In our sample project, we created two mutexes `mux1` and `mux2` to synchronize the access of a shared counter `cnt` by the two threads. After the two threads exit, we destroyed the mutexes.

### Using the mutex

The following four functions are available to lock and unlock a mutex:

```
int pthread_mutex_lock(pthread_mutex_t *mutex);
int pthread_mutex_unlock(pthread_mutex_t *mutex);
int pthread_mutex_trylock(pthread_mutex_t *mutex);
int pthread_mutex_lock_timeout_np(pthread_mutex_t *mutex, unsigned
msecs);
```

In all the four functions, the input argument refers to the `mutex` object in use. A zero return value indicates the mutex is locked or unlocked successfully. The last function allows us to specify a wait timeout in milliseconds. If it cannot acquire the mutex after the timeout, it will return `EBUSY` to indicate failure.

The `pthread_mutex_timedlock` function is defined in some pthread implementations to allow us to specify a timeout value. However, this function is not available in the Android Bionic C library.

We demonstrated the usage of the functions previously in our example. In function `run_by_thread1`, we first lock `mux1` by `pthread_mutex_lock`, and then `mux2` by `pthread_mutex_trylock`. If `mux2` cannot be locked, we unlock `mux1`, sleep for 50 milliseconds, and try again. If `mux2` can be locked, we update the shared counter `cnt`, log its current value, and then release the two mutexes. Another function `run_by_thread2` is similar to `run_by_thread1`, except that it locks `mux2` first, and then `mux1`. The two functions are executed by two threads. This can be illustrated by the following diagram:

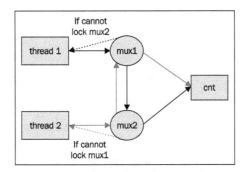

As shown in the preceding diagram, thread 1 needs to obtain `mux1`, then `mux2` in order to update `cnt`, while thread 2 needs to acquire `mux2`, then `mux1` to update `cnt`. In case thread 1 locked `mux1` and thread 2 locked `mux2`, neither threads can proceed. This corresponds to the situation where `pthread_mutex_trylock` returns a nonzero value. If this happens, one thread will give up its mutex so the other thread can proceed to update the shared counter `cnt` and release the two mutexes. Note that we can replace the `pthread_mutex_trylock` with `pthread_mutex_lock_timeout_np` in our code. Readers are encouraged to try it out themselves.

We also implemented a native method `jni_start_threads_dead` which will probably cause a deadlock. The thread setup is similar to the previous case, but we use `pthread_mutex_lock` instead of `pthread_mutex_trylock`, and the threads do not give up the mutexes they have already locked. This can be illustrated as shown in the following diagram:

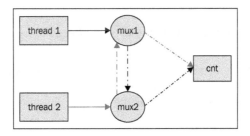

Thread 1 tries to lock `mux1` and then `mux2`, while thread 2 tries to lock `mux2` and then `mux1`. In case where thread 1 has locked `mux1` and thread 2 has locked `mux2`, none of the threads can proceed. Because they won't give up the mutexes they've obtained, the two threads will be blocked forever. This is referred to as a deadlock.

## There's more...

Recall that second input argument for function `pthread_mutex_init` is a pointer to `pthread_mutexattr_t`. A few functions are defined in `pthread.h` to initialize, manipulate, and destroy mutex attributes, including:

- `pthread_mutexattr_init`
- `pthread_mutexattr_destroy`
- `pthread_mutexattr_gettype`
- `pthread_mutexattr_settype`
- `pthread_mutexattr_setpshared`
- `pthread_mutexattr_getpshared`

Interested readers can look into the `pthread.h` header file for more information.

# Synchronizing native threads with conditional variables at Android NDK

The previous recipe discusses how to synchronize threads with mutex. This recipe describes how to use conditional variables.

## How to do it...

The following steps will help us create an Android project that demonstrates the usage of pthread conditional variables:

1. Create an Android application named `NativeThreadsCondVar`. Set the package name as `cookbook.chapter6.nativethreadscondvar`. Refer to the *Loading native libraries and registering native methods* recipe in *Chapter 2, Java Native Interface* for more detailed instructions.

2. Right-click on the project **NativeThreadsCondVar**, select **Android Tools | Add Native Support**.

3. Add a Java file named `MainActivity.java` under the package `cookbook.chapter6.nativethreadscondvar`. This Java file simply loads the native library `NativeThreadsCondVar` and calls the native `jni_start_threads` method.

4. Add two files named `mylog.h` and `NativeThreadsCondVar.cpp` under the `jni` folder. `NativeThreadsCondVar.cpp` contains the code to start two threads. The two threads will update a shared counter. A part of the code is shown as follows:

The `jni_start_threads` function initializes the mutex, conditional variable and creates two threads:

```
pthread_mutex_t mux;
pthread_cond_t cond;
void jni_start_threads() {
  pthread_t th1, th2;
  int threadNum1 = 1, threadNum2 = 2;
  int ret;
  pthread_mutex_init(&mux, NULL);
  pthread_cond_init(&cond, NULL);
  ret = pthread_create(&th1, NULL, run_by_thread1,
void*)&threadNum1);
  LOGI(1, "thread 1 started");
  ret = pthread_create(&th2, NULL, run_by_thread2,
void*)&threadNum2);
  LOGI(1, "thread 2 started");
  ret = pthread_join(th1, NULL);
  LOGI(1, "thread 1 end %d", ret);
  ret = pthread_join(th2, NULL);
  LOGI(1, "thread 2 end %d", ret);
  pthread_mutex_destroy(&mux);
  pthread_cond_destroy(&cond);
}
```

The `run_by_thread1` function is executed by the first native thread:

```
int cnt = 0;
int THR = 10, THR2 = 5;
void *run_by_thread1(void *arg) {
  int* threadNum = (int*)arg;
  pthread_mutex_lock(&mux);
  while (cnt != THR2) {
      LOGI(1, "thread %d: about to wait", *threadNum);
      pthread_cond_wait(&cond, &mux);
  }
  ++cnt;
  LOGI(1, "thread %d: cnt = %d", *threadNum, cnt);
  pthread_mutex_unlock(&mux);
}
```

The `run_by_thread2` function is executed by the second native thread:

```
void *run_by_thread2(void *arg) {
  int* threadNum = (int*)arg;
  while (cnt < THR) {
    pthread_mutex_lock(&mux);
    if (cnt == THR2) {
      pthread_cond_signal(&cond);
    } else {
      ++cnt;
      LOGI(1, "thread %d: cnt = %d", *threadNum, cnt);
    }
    pthread_mutex_unlock(&mux);
    sleep(1);
  }
}
```

5. Add an `Android.mk` file under the `jni` folder with the following content:

```
LOCAL_PATH := $(call my-dir)
include $(CLEAR_VARS)
LOCAL_MODULE    := NativeThreadsCondVar
LOCAL_SRC_FILES := NativeThreadsCondVar.cpp
LOCAL_LDLIBS    := -llog
include $(BUILD_SHARED_LIBRARY)
```

6. Build and run the Android project, and use the following command to monitor the `logcat` output:

```
$ adb logcat -v time NativeThreadsCondVar:I *:S
```

The `logcat` output is shown as follows:

```
10-08 22:35:47.101 I/NativeThreadsCondVar(22060): thread 1 started
10-08 22:35:47.101 I/NativeThreadsCondVar(22060): thread 2 started
10-08 22:35:47.111 I/NativeThreadsCondVar(22060): thread 2: cnt = 1
10-08 22:35:47.111 I/NativeThreadsCondVar(22060): thread 1: about to wait
10-08 22:35:48.112 I/NativeThreadsCondVar(22060): thread 2: cnt = 2
10-08 22:35:49.113 I/NativeThreadsCondVar(22060): thread 2: cnt = 3
10-08 22:35:50.114 I/NativeThreadsCondVar(22060): thread 2: cnt = 4
10-08 22:35:51.115 I/NativeThreadsCondVar(22060): thread 2: cnt = 5
10-08 22:35:52.116 I/NativeThreadsCondVar(22060): thread 1: cnt = 6
10-08 22:35:52.116 I/NativeThreadsCondVar(22060): thread 1 end 0
10-08 22:35:53.117 I/NativeThreadsCondVar(22060): thread 2: cnt = 7
10-08 22:35:54.118 I/NativeThreadsCondVar(22060): thread 2: cnt = 8
10-08 22:35:55.119 I/NativeThreadsCondVar(22060): thread 2: cnt = 9
10-08 22:35:56.120 I/NativeThreadsCondVar(22060): thread 2: cnt = 10
10-08 22:35:57.121 I/NativeThreadsCondVar(22060): thread 2 end 0
```

## How it works...

While mutexes control access of shared data among threads, conditional variables allow threads to synchronize based on the actual value of data. The typical use case is one thread waits for a condition to be satisfied. Without a conditional variable, the thread needs to check for the condition continuously (often known as polling). Conditional variables allow us to handle the situation without the resource consuming polling.

### Initialize and destroy conditional variables

The pthread_cond_init function is used to initialize a conditional variable. It has the following prototype:

```
int pthread_cond_init(pthread_cond_t *cond, const pthread_condattr_t
*attr);
```

The function initializes the conditional variable pointed by the cond input argument with attributes referred by attr argument. If attr is set to NULL, the default attributes are used.

Similar to mutex, a macro PTHREAD_COND_INITIALIZER is defined in pthread.h to initialize a conditional variable with default attributes.

After we are done with the conditional variable, we can destroy it by calling pthread_cond_destroy, which has the following prototype:

```
int pthread_cond_destroy(pthread_cond_t *cond);
```

In our sample code, we called these two functions to initialize and destroy a conditional variable named cond.

### Using the conditional variable:

The following three functions are commonly used to manipulate a conditional variable:

```
int pthread_cond_wait(pthread_cond_t *cond, pthread_mutex_t *mutex);
int pthread_cond_broadcast(pthread_cond_t *cond);
int pthread_cond_signal(pthread_cond_t *cond);
```

All the three functions accept a pointer to the conditional variable in use. The first function also takes a pointer to the associated mutex as the second argument. Note that a conditional variable must be used with an associated mutex.

The first function should be called after the associated mutex is locked; otherwise the function behavior is undefined. It causes the calling thread to block on the conditional variable. In addition, the associated mutex is unlocked automatically and atomically so that another thread can use it.

The second and third functions are used to unblock the threads that were previously blocked on a conditional variable. `pthread_cond_broadcast` will unblock all threads that are blocked on the conditional variable pointed by `cond`, while `pthread_cond_signal` will unblock at least one of the threads blocked on `cond`. The two functions have no effect if no threads are blocked on the conditional variable specified by `cond`. In case there are multiple threads to unblock, the order is dependent on the scheduling policy, which we will discuss in the *Scheduling native threads at Android NDK* recipe later in this chapter.

The usage of these functions is demonstrated in our sample code. In the `run_by_thread1` function, thread one will lock the associated mutex, and then wait on the conditional variable `cond`. This will cause thread one to release the mutex `mux`. In function `run_by_thread2`, thread two will obtain `mux` and increase the shared counter `cnt`.

When `cnt` is increased to five, thread two calls `pthread_cond_signal` to unblock thread one and release `mux`. Thread one will lock `mux` automatically and atomically (note that no `pthread_mutex_lock` call is needed upon wake up), and then increase `cnt` from five to six, and finally exit. Thread two will continue to increase the `cnt` value to 10 and exit. This explains the preceding screenshot.

> We put the `pthread_cond_wait(&cond, &mux)` function inside a while loop to handle spurious wakeup. Spurious wakeup refers to the case where a thread is woken up even though no thread signaled the condition. It is recommended that we always check the condition when `pthread_cond_wait` is returned. You can refer to `http://pubs.opengroup.org/onlinepubs/7908799/xsh/pthread_cond_wait.html` for more information.

## There's more...

The sample project demonstrates how conditional variables are used for native threads synchronization. We will go through the details in the following section.

### Conditional variable attributes functions

In our sample code, we created the conditional variable with default attributes by specifying the second argument to `pthread_cond_init` as `NULL`. `pthread.h` defines a few functions to initialize and manipulate conditional variable attributes. These functions include `pthread_condattr_init`, `pthread_condattr_getpshared`, `pthread_condattr_setpshared`, and `pthread_condattr_destroy`. We will not discuss these functions because they are not used often. Interested readers can refer to the `pthread.h` header file available at `platforms/android-<API level>/arch-arm/usr/include/` for more information.

## Timed conditional variable functions

`pthread.h` also defines a few functions that allow us to specify a timeout value for waiting on a conditional variable. They are listed as follows:

```
int pthread_cond_timedwait(pthread_cond_t *cond, pthread_mutex_t *
mutex, const struct timespec *abstime);
int pthread_cond_timedwait_monotonic_np(pthread_cond_t *cond, pthread_
mutex_t        *mutex, const struct timespec  *abstime);
int pthread_cond_timedwait_relative_np(pthread_cond_t *cond, pthread_
mutex_t        *mutex, const struct timespec  *reltime);
int pthread_cond_timeout_np(pthread_cond_t *cond, pthread_mutex_t *
mutex, unsigned msecs);
```

The first two functions `pthread_cond_timedwait` and `pthread_cond_timedwait_monotonic_np` allow us to specify an absolute time value. When the system time equals or exceeds the specified time, a timeout error is returned. The difference between the two functions is that the first function uses the wall clock while the second function uses the `CLOCK_MONOTONIC` clock. The system wall clock can jump forwards or backwards (for example, the wall clock of a machine configured to use Network Time Protocol may change upon clock synchronization), while the `CLOCK_MONOTONIC` clock is the absolute time elapsed since some fixed point in the past and it cannot be changed abruptly.

> Android `pthread.h` also defines a function `pthread_cond_timedwait_monotonic`, which is deprecated. It is functionally equivalent to `pthread_cond_timedwait_monotonic_np`. We should always use `pthread_cond_timedwait_monotonic_np` instead.

The last two functions `pthread_cond_timedwait_relative_np` and `pthread_cond_timeout_np` allow us to specify a relative timeout value with respect to the current time. The difference is that the timeout value is specified as `timespec` structure in one function and as number of milliseconds in the other.

> Several methods covered in this recipe end with `np`, which stands for "nonportable". This means these functions may not be implemented in other pthread libraries. If we are designing our program to also work on platforms other than Android, we should avoid using these functions.

# Synchronizing native threads with reader/ writer locks at Android NDK

The previous two recipes cover thread synchronization with mutex and conditional variables. This recipe discusses reader/writer locks in Android NDK.

## Getting ready...

Readers are recommended to read the previous two recipes, *Synchronizing native threads with mutex at Android NDK* and *Synchronizing native threads with conditional variables at Android NDK*, before going through this one.

## How to do it...

The following steps will help you create an Android project that demonstrates the usage of the pthread reader/writer lock:

1.  Create an Android application named `NativeThreadsRWLock`. Set the package name as `cookbook.chapter6.nativethreadsrwlock`. Refer to the *Loading native libraries and registering native methods* recipe in *Chapter 2, Java Native Interface* for more detailed instructions.

2.  Right-click on the project **NativeThreadsRWLock**, select **Android Tools | Add Native Support**.

3.  Add a Java file named `MainActivity.java` under package `cookbook. chapter6.nativethreadsrwlock`. This Java file simply loads the native library `NativeThreadsRWLock` and calls the native method `jni_start_threads`.

4.  Add two files named `mylog.h` and `NativeThreadsRWLock.cpp` under the `jni` folder. A part of the code in `NativeThreadsRWLock.cpp` is shown as follows:

    `jni_start_threads` starts `pNumOfReader` reader threads and `pNumOfWriter` writer threads:

```
void jni_start_threads(JNIEnv *pEnv, jobject pObj, int
pNumOfReader, int pNumOfWriter) {
  pthread_t *ths;
  int i, ret;
  int *thNum;
  ths = (pthread_t*)malloc(sizeof(pthread_t)*(pNumOfReader+pNumOfW
riter));
  thNum = (int*)malloc(sizeof(int)*(pNumOfReader+pNumOfWriter));
  pthread_rwlock_init(&rwlock, NULL);
  for (i = 0; i < pNumOfReader + pNumOfWriter; ++i) {
    thNum[i] = i;
```

```
    if (i < pNumOfReader) {
        ret = pthread_create(&ths[i], NULL, run_by_read_thread,
(void*)&(thNum[i]));
    } else {
        ret = pthread_create(&ths[i], NULL, run_by_write_thread,
(void*)&(thNum[i]));
    }
  }
  for (i = 0; i < pNumOfReader+pNumOfWriter; ++i) {
    ret = pthread_join(ths[i], NULL);
  }
  pthread_rwlock_destroy(&rwlock);
  free(thNum);
  free(ths);
}
```

The `run_by_read_thread` function is executed by the reader threads:

```
void *run_by_read_thread(void *arg) {
  int* threadNum = (int*)arg;
  int ifRun = 1;
  int accessTimes = 0;
  int ifPrint = 1;
  while (ifRun) {
    if (!pthread_rwlock_rdlock(&rwlock)) {
      if (100000*numOfWriter == sharedCnt) {
        ifRun = 0;
      }
      if (0 <= sharedCnt && ifPrint) {
        LOGI(1, "reader thread %d sharedCnt value before
processing %d\n", *threadNum, sharedCnt);
        int j, k;//some dummy processing
        for (j = 0; j < 100000; ++j) {
          k = j*2;
          k = sqrt(k);
        }
        ifPrint = 0;
        LOGI(1, "reader thread %d sharedCnt value after processing
%d %d\n", *threadNum, sharedCnt, k);
      }
      if ((++accessTimes) == INT_MAX/5) {
        accessTimes = 0;
        LOGI(1, "reader thread %d still running: %d\n",
*threadNum, sharedCnt);
      }
      pthread_rwlock_unlock(&rwlock);
```

```
        }
    }
    LOGI(1, "reader thread %d return %d\n", *threadNum, sharedCnt);
    return NULL;
}
```

The `run_by_write_thread` function is executed by the writer threads:

```
void *run_by_write_thread(void *arg) {
    int cnt = 100000, i, j, k;
    int* threadNum = (int*)arg;
    for (i = 0; i < cnt; ++i) {
        if (!pthread_rwlock_wrlock(&rwlock)) {
            int lastShCnt = sharedCnt;
            for (j = 0; j < 10; ++j) {   //some dummy processing
                k = j*2;
                k = sqrt(k);
            }
            sharedCnt = lastShCnt + 1;
            pthread_rwlock_unlock(&rwlock);
        }
    }
    LOGI(1, "writer thread %d return %d %d\n", *threadNum,
sharedCnt, k);
    return NULL;
}
```

5. Add an `Android.mk` file under the `jni` folder with the following content:

```
LOCAL_PATH := $(call my-dir)
include $(CLEAR_VARS)
LOCAL_MODULE    := NativeThreadsRWLock
LOCAL_SRC_FILES := NativeThreadsRWLock.cpp
LOCAL_LDLIBS    := -llog
include $(BUILD_SHARED_LIBRARY)
```

6. Build and run the Android project, and use the following command to monitor the `logcat` output:

```
$ adb logcat -v time NativeThreadsRWLock:I *:S
```

The `logcat` output is shown as follows:

```
10-11 21:26:34.864 I/NativeThreadsRWLock(18213): readers: 2, writers: 3
10-11 21:26:34.864 I/NativeThreadsRWLock(18213): reader thread 0 started
10-11 21:26:34.864 I/NativeThreadsRWLock(18213): reader thread 1 started
10-11 21:26:34.864 I/NativeThreadsRWLock(18213): writer thread 2 started
10-11 21:26:34.874 I/NativeThreadsRWLock(18213): writer thread 3 started
10-11 21:26:34.874 I/NativeThreadsRWLock(18213): writer thread 4 started
10-11 21:26:34.904 I/NativeThreadsRWLock(18213): reader thread 1 sharedCnt value before processing 2261
10-11 21:26:34.904 I/NativeThreadsRWLock(18213): reader thread 0 sharedCnt value before processing 2261
10-11 21:26:35.124 I/NativeThreadsRWLock(18213): reader thread 1 sharedCnt value after processing 2261 447
10-11 21:26:35.134 I/NativeThreadsRWLock(18213): reader thread 0 sharedCnt value after processing 2261 447
10-11 21:26:38.798 I/NativeThreadsRWLock(18213): writer thread 2 return 290142 4
10-11 21:26:38.948 I/NativeThreadsRWLock(18213): writer thread 3 return 297538 4
10-11 21:26:38.998 I/NativeThreadsRWLock(18213): writer thread 4 return 300000 4
10-11 21:26:38.998 I/NativeThreadsRWLock(18213): reader thread 1 return 300000
10-11 21:26:38.998 I/NativeThreadsRWLock(18213): reader thread 0 return 300000
```

## How it works...

The reader/writer lock is internally implemented with a mutex and a conditional variable. It has the following rules:

- ▶ If a thread tries to acquire a read lock for a resource, it can succeed as long as no other threads hold a write lock for the resource.

- ▶ If a thread tries to acquire a write lock for a resource, it can succeed only when no other threads hold a write or read lock for the resource.

- ▶ The reader/writer lock guarantees only one thread can modify (need to get the write lock) the resource, while permitting multiple threads to read the resource (need to get the read lock). It also makes sure no reads happen when the resource is being changed. In the following sections we describe the reader/writer lock functions provided by Android `pthread.h`.

### Initialize and destroy a reader/writer lock

The following two functions are defined to initialize and destroy a reader/writer lock:

```
int pthread_rwlock_init(pthread_rwlock_t *rwlock, const pthread_
rwlockattr_t *attr);
int pthread_rwlock_destroy(pthread_rwlock_t *rwlock);
```

`pthread_rwlock_init` initializes a reader/writer lock pointed by the `rwlock` argument with the attributes referred by argument `attr`. If `attr` is set to `NULL`, the default attributes are used. `pthread_rwlock_destroy` accepts a pointer to a reader/writer lock and destroys it.

There is also a macro `PTHREAD_RWLOCK_INITIALIZER` defined to initialize a reader/writer lock. The default attributes are used in this case.

## Using a reader/writer lock

The following two functions are defined to acquire a read and a write lock respectively:

```
int pthread_rwlock_rdlock(pthread_rwlock_t *rwlock);
int pthread_rwlock_wrlock(pthread_rwlock_t *rwlock);
```

Both functions accept a pointer to the reader/writer lock and return a zero to indicate success. If the lock cannot be acquired, the calling thread will be blocked until the block is available or till an error occurs.

The following function is defined to unlock either read lock or write lock:

```
int pthread_rwlock_unlock(pthread_rwlock_t *rwlock);
```

In our sample code, we demonstrated the usage of these functions. In `run_by_read_thread` function, the read threads need to acquire the read lock in order to access the value of the shared resource `sharedCnt`. In the `run_by_write_thread` function, the write threads need to acquire the write lock before updating the shared resource `sharedCnt`.

If we remove the code which locks and unlocks the read and write lock, build the application, and rerun it, the output is as shown in the following screenshot:

```
10-11 21:24:12.985 I/NativeThreadsRWLock(18155): readers: 2, writers: 3
10-11 21:24:12.985 I/NativeThreadsRWLock(18155): reader thread 0 started
10-11 21:24:12.985 I/NativeThreadsRWLock(18155): reader thread 1 started
10-11 21:24:12.985 I/NativeThreadsRWLock(18155): writer thread 2 started
10-11 21:24:12.985 I/NativeThreadsRWLock(18155): writer thread 3 started
10-11 21:24:12.985 I/NativeThreadsRWLock(18155): writer thread 4 started
10-11 21:24:13.005 I/NativeThreadsRWLock(18155): reader thread 1 sharedCnt value before processing 1734
10-11 21:24:13.015 I/NativeThreadsRWLock(18155): reader thread 0 sharedCnt value before processing 1734
10-11 21:24:13.526 I/NativeThreadsRWLock(18155): reader thread 1 sharedCnt value after processing 11788 447
10-11 21:24:13.536 I/NativeThreadsRWLock(18155): reader thread 0 sharedCnt value after processing 12565 447
10-11 21:24:16.959 I/NativeThreadsRWLock(18155): writer thread 3 return 100000 4
10-11 21:24:16.989 I/NativeThreadsRWLock(18155): writer thread 2 return 100487 4
10-11 21:24:16.999 I/NativeThreadsRWLock(18155): writer thread 4 return 101340 4
10-11 21:24:29.772 I/NativeThreadsRWLock(18155): reader thread 0 still running: 13876
10-11 21:24:29.782 I/NativeThreadsRWLock(18155): reader thread 1 still running: 12446
10-11 21:24:43.856 I/NativeThreadsRWLock(18155): reader thread 1 still running: 101340
10-11 21:24:43.866 I/NativeThreadsRWLock(18155): reader thread 0 still running: 101340
```

As shown in the output, the shared resource `sharedCnt` is updated to a value less than the final value when reader/writer lock is enabled. The reason is illustrated in the following diagram:

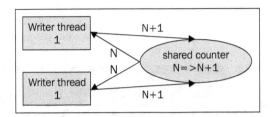

In this diagram, two writers get the same value (N) of the shared counter, and both update the value from N to N+1. When they write the value back to the shared counter, the shared counter is updated from N to N+1 although it is updated twice by two writers. This illustrates why we need write lock. Also note at reader threads, two reads of the `sharedCnt` (one before processing and one after processing) give us two different values because the writers have updated the value. This may not be desirable sometimes and that is why a read lock is necessary at times.

## There's more...

There are a few more read/write lock functions defined in `pthread.h`.

### Timed read/write lock and trylock

Android `pthread.h` defines the following two functions to allow the calling thread to specify a timeout value when trying to acquire the read or write lock:

```
int pthread_rwlock_timedrdlock(pthread_rwlock_t *rwlock, const struct
timespec *abs_timeout);
int pthread_rwlock_timedwrlock(pthread_rwlock_t *rwlock, const struct
timespec *abs_timeout);
```

In addition, the following two functions are available for the calling thread to acquire read or write lock without blocking itself. If the lock is not available, the functions will return a nonzero value instead of blocking:

```
int pthread_rwlock_tryrdlock(pthread_rwlock_t *rwlock);
int pthread_rwlock_trywrlock(pthread_rwlock_t *rwlock);
```

## Reader/writer lock attribute functions

Android `pthread.h` defines a set of functions to initialize and manipulate a reader/writer lock attribute, which can be passed to `pthread_rwlock_init` as the second argument. These functions include `pthread_rwlockattr_init`, `pthread_rwlockattr_destroy`, `pthread_rwlockattr_setpshared`, and `pthread_rwlockattr_getpshared`. They are not used often in Android NDK development and therefore not discussed here.

# Synchronizing native threads with semaphore at Android NDK

We have covered mutex, conditional variables, and reader/writer lock in the previous three recipes. This is the last recipe on threads synchronization at Android NDK, and we will discuss semaphores.

## Getting ready...

Readers are expected to read through the previous three recipes, *Synchronizing native threads with mutex at Android NDK*, *Synchronizing native threads with conditional variables at Android NDK*, and *Synchronizing native threads with reader/writer locks at Android NDK*, before this one.

## How to do it...

The following steps will help you create an Android project that demonstrates the usage of pthread reader/writer lock:

1. Create an Android application named `NativeThreadsSemaphore`. Set the package name as `cookbook.chapter6.nativethreadssemaphore`. Refer to the *Loading native libraries and registering native methods* recipe in *Chapter 2, Java Native Interface* for more detailed instructions.

2. Right-click on the project **NativeThreadsSemaphore**, select **Android Tools | Add Native Support**.

3. Add a Java file named `MainActivity.java` under package `cookbook.chapter6.nativethreadssemaphore`. This Java file simply loads the native library `NativeThreadsSemaphore` and calls the native `jni_start_threads` method.

4. Add two files named `mylog.h` and `NativeThreadsSemaphore.cpp` under the `jni` folder. A part of the code in `NativeThreadsSemaphore.cpp` is shown as follows:

`jni_start_threads` creates `pNumOfConsumer` number of consumer threads, `pNumOfProducer` number of producer threads, and `numOfSlots` number of slots:

```
void jni_start_threads(JNIEnv *pEnv, jobject pObj, int
pNumOfConsumer, int pNumOfProducer, int numOfSlots) {
  pthread_t *ths;
  int i, ret;
  int *thNum;
  pthread_mutex_init(&mux, NULL);
  sem_init(&emptySem, 0, numOfSlots);
  sem_init(&fullSem, 0, 0);
  ths = (pthread_t*)malloc(sizeof(pthread_t)*(pNumOfConsumer+pNumO
fProducer));
  thNum = (int*)malloc(sizeof(int)*(pNumOfConsumer+pNumOfProduc
er));
  for (i = 0; i < pNumOfConsumer + pNumOfProducer; ++i) {
    thNum[i] = i;
    if (i < pNumOfConsumer) {
      ret = pthread_create(&ths[i], NULL,
un_by_consumer_thread, (void*)&(thNum[i]));
    } else {
      ret = pthread_create(&ths[i], NULL, run_by_producer_thread,
(void*)&(thNum[i]));
    }
  }
  for (i = 0; i < pNumOfConsumer+pNumOfProducer; ++i) {
    ret = pthread_join(ths[i], NULL);
  }
  sem_destroy(&emptySem);
  sem_destroy(&fullSem);
  pthread_mutex_destroy(&mux);
  free(thNum);
  free(ths);
}
```

`run_by_consumer_thread` is the function executed by the consumer thread:

```
void *run_by_consumer_thread(void *arg) {
  int* threadNum = (int*)arg;
  int i;
  for (i = 0; i < 4; ++i) {
    sem_wait(&fullSem);
    pthread_mutex_lock(&mux);
    --numOfItems;
```

```
        pthread_mutex_unlock(&mux);
        sem_post(&emptySem);
    }
    return NULL;
}
```

run_by_producer_thread is the function executed by producer thread:

```
void *run_by_producer_thread(void *arg) {
    int* threadNum = (int*)arg;
    int i;
    for (i = 0; i < 4; ++i) {
        sem_wait(&emptySem);
        pthread_mutex_lock(&mux);
        ++numOfItems;
        pthread_mutex_unlock(&mux);
        sem_post(&fullSem);
    }
    return NULL;
}
```

5. Add an `Android.mk` file under the `jni` folder with the following content:

```
LOCAL_PATH := $(call my-dir)
include $(CLEAR_VARS)
LOCAL_MODULE    := NativeThreadsSemaphore
LOCAL_SRC_FILES := NativeThreadsSemaphore.cpp
LOCAL_LDLIBS    := -llog
include $(BUILD_SHARED_LIBRARY)
```

6. Build and run the Android project, and use the following command to monitor the `logcat` output:

```
$ adb logcat -v time NativeThreadsSemaphore:I *:S
```

The `logcat` output is shown in the following screenshot:

```
10-11 22:15:34.332 I/NativeThreadsSemaphore(21349): consumer thread 0 started
10-11 22:15:34.332 I/NativeThreadsSemaphore(21349): consumer thread 1 started
10-11 22:15:34.332 I/NativeThreadsSemaphore(21349): producer thread 2 started
10-11 22:15:34.332 I/NativeThreadsSemaphore(21349): producer thread 3 started
10-11 22:15:34.342 I/NativeThreadsSemaphore(21349): producer thread 2 produced 1
10-11 22:15:34.342 I/NativeThreadsSemaphore(21349): producer thread 2 produced 2
10-11 22:15:34.342 I/NativeThreadsSemaphore(21349): producer thread 2 produced 3
10-11 22:15:34.342 I/NativeThreadsSemaphore(21349): consumer thread 1 consumed 3
10-11 22:15:34.342 I/NativeThreadsSemaphore(21349): consumer thread 1 consumed 2
10-11 22:15:34.342 I/NativeThreadsSemaphore(21349): consumer thread 1 consumed 1
10-11 22:15:34.342 I/NativeThreadsSemaphore(21349): producer thread 3 produced 1
10-11 22:15:34.342 I/NativeThreadsSemaphore(21349): producer thread 3 produced 2
10-11 22:15:34.342 I/NativeThreadsSemaphore(21349): producer thread 3 produced 3
10-11 22:15:34.352 I/NativeThreadsSemaphore(21349): consumer thread 1 consumed 3
10-11 22:15:34.352 I/NativeThreadsSemaphore(21349): consumer thread 1 return
10-11 22:15:34.352 I/NativeThreadsSemaphore(21349): producer thread 2 produced 3
10-11 22:15:34.352 I/NativeThreadsSemaphore(21349): producer thread 2 return
10-11 22:15:34.352 I/NativeThreadsSemaphore(21349): consumer thread 0 consumed 3
10-11 22:15:34.352 I/NativeThreadsSemaphore(21349): consumer thread 0 consumed 2
10-11 22:15:34.352 I/NativeThreadsSemaphore(21349): consumer thread 0 consumed 1
10-11 22:15:34.352 I/NativeThreadsSemaphore(21349): producer thread 3 produced 1
10-11 22:15:34.352 I/NativeThreadsSemaphore(21349): producer thread 3 return
10-11 22:15:34.352 I/NativeThreadsSemaphore(21349): consumer thread 0 consumed 1
10-11 22:15:34.352 I/NativeThreadsSemaphore(21349): consumer thread 0 return
```

## How it works...

Semaphores are essentially integer counters. Two primary operations are supported by a semaphore:

- **Wait**: It attempts to decrement the semaphore value. If wait is called on a semaphore of value zero, the calling thread is blocked until `post` is called somewhere else to increment semaphore value.

- **Post**: It attempts to increment the semaphore value.

The semaphore related functions are defined in `semaphore.h` rather than `pthread.h`. Next, we describe a few key functions.

>  Interprocess mutex, conditional variable, and semaphore are not supported on Android. Android uses `Intent`, `Binder`, and so on for interprocess communication and synchronization.

## Initialize and destroy a semaphore

The following three functions are defined to initialize or destroy a semaphore:

```
extern int sem_init(sem_t *sem, int pshared, unsigned int value);
extern int    sem_init(sem_t *, int, unsigned int value);
extern int    sem_destroy(sem_t *);
```

The first two functions are used to initialize a semaphore. They both initialize the semaphore pointed by the input argument sem with the value indicated by the argument value. The first function also accepts an argument pshared, which should be set to zero for thread synchronization. If it is set to nonzero, the semaphore can be shared between processes, which is not supported on Android and therefore not discussed.

## Using a semaphore

The following functions are defined to use a semaphore.

```
extern int    sem_trywait(sem_t *);
extern int    sem_wait(sem_t *);
extern int    sem_post(sem_t *);
extern int    sem_getvalue(sem_t *, int *);
```

The first two functions are used to wait on a semaphore. If the semaphore value is not zero, then the value is decreased by one. If the value is zero, the first function will return a nonzero value to indicate failure, while the second function will block the calling thread. The third function is used to increase the semaphore value by one, and the last function is used to query the value of the semaphore. Note that the value is returned through the second input argument rather than the return value.

 Android semaphore.h also defines a function named sem_timedwait to allow us to specify a timeout value while waiting on a semaphore.

In our sample project, we used two semaphores emptySem and fullSem, and a mutex mux. The app will create a few producer threads and consumer threads. The emptySem semaphore is used to indicate the number of slots available to store the items produced by the producer thread, while fullSem refers to the number of items for the consumer thread to consume. The mutex mux is used to ensure no two threads can access the shared counter numOfItems at one time.

The producer thread will need to wait on the emptySem semaphore. When it is unblocked, the producer has obtained an empty slot. It will lock mux and then update the shared count numOfItems, which means a new item has been produced. Therefore, it will call the post function on fullSem to increment its value.

On the other hand, the consumer thread will wait on `fullSem`. When it is unblocked, the consumer has consumed an item. It will lock `mux` and then update the shared count `numOfItems`. A new empty slot is available because of the item consumed, so the consumer thread will call post on `emptySem` to increment its value.

 Mutex `mux` can also be replaced by a binary semaphore. The possible values of a binary semaphore are restricted to zero and one.

# Scheduling native threads at Android NDK

This recipe discusses how to schedule native threads at Android NDK.

## Getting ready...

Readers are suggested to read the *Manipulating classes in JNI* and *Calling static and instance methods from native code* recipes in *Chapter 2, Java Native Interface*, and *Creating and terminating native threads at Android NDK* recipe in this chapter.

## How to do it...

The following steps will help us create an Android project that demonstrates threads scheduling at Android NDK:

1.  Create an Android application named `NativeThreadsSchedule`. Set the package name as `cookbook.chapter6.nativethreadsschedule`. Refer to the *Loading native libraries and registering native methods* recipe in *Chapter 2, Java Native Interface* for more detailed instructions.

2.  Right-click on the project **NativeThreadsSchedule**, select **Android Tools | Add Native Support**.

3.  Add a Java file named `MainActivity.java` under package `cookbook.chapter6.nativethreadsschedule`. This Java file simply loads the native library `NativeThreadsSchedule` and calls the native methods.

4.  Add five files named `mylog.h`, `NativeThreadsSchedule.h`, `NativeThreadsSchedule.cpp`, `SetPriority.cpp`, and `JNIProcessSetThreadPriority.cpp` under the `jni` folder. A part of the code in the last three files is shown as follows:

    □   The `NativeThreadsSchedule.cpp` file contains the source code to demonstrate the threads scheduling functions defined in `pthread.h` `jni_thread_scope` demonstrates how to set the native thread contention scope:

```
void jni_thread_scope() {
  pthread_attr_t attr;
  int ret;
  pid_t fpId = fork();
  if (0 == fpId) {
    pthread_attr_init(&attr);
    int ret = pthread_attr_setscope(&attr, PTHREAD_SCOPE_
PROCESS);
    pthread_t thFive[5];
    int threadNums[5];
    int i;
    for (i = 0; i < 5; ++i) {
      threadNums[i] = i;        ret = pthread_
create(&thFive[i], &attr, run_by_thread,
(void*)&(threadNums[i]));
    }
    for (i = 0; i < 5; ++i) {
      ret = pthread_join(thFive[i], NULL);
    }
  } else {
    pthread_attr_init(&attr);
    int ret = pthread_attr_setscope(&attr, PTHREAD_SCOPE_
SYSTEM);
    pthread_t th1;
    int threadNum1 = 0;
    ret = pthread_create(&th1, &attr, run_by_thread,
(void*)&threadNum1);
    ret = pthread_join(th1, NULL);
  }
  //code executed by both processes
  pthread_attr_destroy(&attr);
}
```

`jni_thread_fifo` demonstrates how to set the native thread scheduling policy and priority:

```
void jni_thread_fifo() {
  pthread_attr_t attr;
  int ret;
  pid_t fpId = fork();
  struct sched_param prio;
  if (0 == fpId) {
    //the child process
```

```
      pthread_attr_init(&attr);
      pthread_t thFive[5];
      int threadNums[5];
      int i;
      for (i = 0; i < 5; ++i) {
        if (i == 4) {
          prio.sched_priority = 10;
        } else {
          prio.sched_priority = 1;
        }
        ret = pthread_attr_setschedpolicy(&attr, SCHED_FIFO);
        ret = pthread_attr_setschedparam(&attr, &prio);
        threadNums[i] = i;
        ret = pthread_create(&thFive[i], &attr, run_by_thread,
(void*)&(threadNums[i]));
        pthread_attr_t lattr;
        struct sched_param lprio;
        int lpolicy;
        pthread_getattr_np(thFive[i], &lattr);
        pthread_attr_getschedpolicy(&lattr, &lpolicy);
        pthread_attr_getschedparam(&lattr, &lprio);
        pthread_attr_destroy(&lattr);
      }
      for (i = 0; i < 5; ++i) {
        ret = pthread_join(thFive[i], NULL);
      }
  } else {
    //the parent process
    pthread_attr_init(&attr);
    prio.sched_priority = 10;
    ret = pthread_attr_setschedpolicy(&attr, SCHED_FIFO);
    ret = pthread_attr_setschedparam(&attr, &prio);
    pthread_t th1;
    int threadNum1 = 0;
    ret = pthread_create(&th1, &attr, run_by_thread,
(void*)&threadNum1);
    pthread_attr_t lattr;
    struct sched_param lprio;
    int lpolicy;
    pthread_getattr_np(th1, &lattr);
    pthread_attr_getschedpolicy(&lattr, &lpolicy);
    pthread_attr_getschedparam(&lattr, &lprio);
    pthread_attr_destroy(&lattr);
    ret = pthread_join(th1, NULL);
```

```
    }
    //code executed by both processes
    pthread_attr_destroy(&attr);
}
```

`run_by_thread` is the actual function to be executed by each native thread:

```
void *run_by_thread(void *arg) {
    int cnt = 18000000, i;
    int* threadNum = (int*)arg;
    for (i = 1; i < cnt; ++i) {
        if (0 == i%6000000) {
            LOGI(1, "process %d thread %d: %d", getpid(),
*threadNum, i);
        }
    }
    LOGI(1, "process %d thread %d return", getpid(),
*threadNum);
}
```

- ❑ The `SetPriority.cpp` file contains the source code to configure thread nice value through `setpriority`

    The `jni_thread_set_priority` method creates and joins five native methods:

    ```
    void jni_thread_set_priority() {
        int ret;
        pthread_t thFive[5];
        int threadNums[5];
        int i;
        for (i = 0; i < 5; ++i) {
            threadNums[i] = i;
            ret = pthread_create(&thFive[i], NULL, run_by_thread2,
    (void*)&(threadNums[i]));
        }
        for (i = 0; i < 5; ++i) {
            ret = pthread_join(thFive[i], NULL);
        }
    }
    ```

    The `run_by_thread2` function is executed by each native thread:

    ```
    void *run_by_thread2(void *arg) {
        int cnt = 18000000, i;
        int* threadNum = (int*)arg;
        switch (*threadNum) {
    ```

```
    case 0:
      setpriority(PRIO_PROCESS, 0, 21);
      break;
    case 1:
      setpriority(PRIO_PROCESS, 0, 10);
      break;
    case 2:
      setpriority(PRIO_PROCESS, 0, 0);
      break;
    case 3:
      setpriority(PRIO_PROCESS, 0, -10);
      break;
    case 4:
      setpriority(PRIO_PROCESS, 0, -21);
      break;
    default:
      break;
    }
    for (i = 1; i < cnt; ++i) {
      if (0 == i%6000000) {
        int prio = getpriority(PRIO_PROCESS, 0);
        LOGI(1, "thread %d (prio = %d): %d", *threadNum, prio,
i);
      }
    }
    int prio = getpriority(PRIO_PROCESS, 0);
    LOGI(1, "thread %d (prio = %d): %d return", *threadNum,
prio, i);
    }
```

❏   The JNIProcessSetThreadPriority.cpp file contains the source
    code to configure thread nice value through the android.os.Process.
    setThreadPriority Java method

The jni_process_setThreadPriority method creates and joins five
native threads:

```
void jni_process_setThreadPriority() {
  int ret;
  pthread_t thFive[5];
  int threadNums[5];
  int i;
  for (i = 0; i < 5; ++i) {
    threadNums[i] = i;
    ret = pthread_create(&thFive[i], NULL, run_by_thread3,
(void*)&(threadNums[i]));
```

```
      if(ret) {
        LOGE(1, "cannot create the thread %d: %d", i, ret);
      }
      LOGI(1, "thread %d started", i);
    }
  for (i = 0; i < 5; ++i) {
    ret = pthread_join(thFive[i], NULL);
    LOGI(1, "join returned for thread %d", i);
  }
}
```

The `run_by_thread3` function is executed by each native thread. The thread nice value is set here:

```
void *run_by_thread3(void *arg) {
  int cnt = 18000000, i;
  int* threadNum = (int*)arg;
  JNIEnv *env;
  jmethodID setThreadPriorityMID;
  cachedJvm->AttachCurrentThread(&env, NULL);
  jclass procClass = env->FindClass("android/os/Process");
  setThreadPriorityMID = env->GetStaticMethodID(procClass,
"setThreadPriority", "(I)V");
  switch (*threadNum) {
  case 0:
    env->CallStaticVoidMethod(procClass,
setThreadPriorityMID, 21);
    break;
  case 1:
    env->CallStaticVoidMethod(procClass,
setThreadPriorityMID, 10);
    break;
  case 2:
    env->CallStaticVoidMethod(procClass,
setThreadPriorityMID, 0);
    break;
  case 3:
    env->CallStaticVoidMethod(procClass,
setThreadPriorityMID, -10);
    break;
  case 4:
    env->CallStaticVoidMethod(procClass,
setThreadPriorityMID, -21);
    break;
  default:
    break;

  }
  //we can also use getThreadPriority(int tid) through JNI
interface
```

```
for (i = 1; i < cnt; ++i) {
  if (0 == i%6000000) {
      int prio = getpriority(PRIO_PROCESS, 0);
      LOGI(1, "thread %d (prio = %d): %d", *threadNum, prio,
i);
  }
}
int prio = getpriority(PRIO_PROCESS, 0);
  LOGI(1, "thread %d (prio = %d): %d return", *threadNum,
prio, i);
  cachedJvm->DetachCurrentThread();
}
```

5. Add an `Android.mk` file under the `jni` folder with the following content:

```
LOCAL_PATH := $(call my-dir)
include $(CLEAR_VARS)
LOCAL_MODULE     := NativeThreadsSchedule
LOCAL_SRC_FILES := NativeThreadsSchedule.cpp
LOCAL_LDLIBS     := -llog
include $(BUILD_SHARED_LIBRARY)
```

6. In `MainActivity.java`, disable all native methods except `jni_thread_scope`. Build the project and run it. Start a terminal and enter the following command to monitor the `logcat` output:

```
$ adb logcat -v time NativeThreadsSchedule:I *:S
```

The following screenshot shows the output:

```
10-10 23:12:05.503 I/NativeThreadsSchedule(  688): process 688 set scope 0 0
10-10 23:12:05.503 I/NativeThreadsSchedule(  688): process 688 thread 0 started
10-10 23:12:05.513 I/NativeThreadsSchedule(  702): process 702 set scope 95 0
10-10 23:12:05.513 I/NativeThreadsSchedule(  702): process 702 thread 0 started
10-10 23:12:05.513 I/NativeThreadsSchedule(  702): process 702 thread 1 started
10-10 23:12:05.513 I/NativeThreadsSchedule(  702): process 702 thread 2 started
10-10 23:12:05.523 I/NativeThreadsSchedule(  702): process 702 thread 3 started
10-10 23:12:05.523 I/NativeThreadsSchedule(  702): process 702 thread 4 started
10-10 23:12:10.122 I/NativeThreadsSchedule(  702): process 702 thread 0: 6000000
10-10 23:12:10.122 I/NativeThreadsSchedule(  702): process 702 thread 1: 6000000
10-10 23:12:10.152 I/NativeThreadsSchedule(  702): process 702 thread 2: 6000000
10-10 23:12:10.162 I/NativeThreadsSchedule(  688): process 688 thread 0: 6000000
10-10 23:12:10.213 I/NativeThreadsSchedule(  702): process 702 thread 3: 6000000
10-10 23:12:10.333 I/NativeThreadsSchedule(  702): process 702 thread 4: 6000000
10-10 23:12:16.563 I/NativeThreadsSchedule(  702): process 702 thread 0: 12000000
10-10 23:12:16.663 I/NativeThreadsSchedule(  702): process 702 thread 1: 12000000
10-10 23:12:16.683 I/NativeThreadsSchedule(  688): process 688 thread 0: 12000000
10-10 23:12:16.743 I/NativeThreadsSchedule(  702): process 702 thread 2: 12000000
10-10 23:12:16.773 I/NativeThreadsSchedule(  702): process 702 thread 3: 12000000
10-10 23:12:16.893 I/NativeThreadsSchedule(  702): process 702 thread 4: 12000000
10-10 23:12:23.062 I/NativeThreadsSchedule(  702): process 702 thread 0 return
10-10 23:12:23.112 I/NativeThreadsSchedule(  702): process 702 thread 1 return
10-10 23:12:23.153 I/NativeThreadsSchedule(  688): process 688 thread 0 return
10-10 23:12:23.192 I/NativeThreadsSchedule(  702): process 702 thread 2 return
10-10 23:12:23.213 I/NativeThreadsSchedule(  702): process 702 thread 3 return
10-10 23:12:23.304 I/NativeThreadsSchedule(  702): process 702 thread 4 return
```

7. In `MainActivity.java`, disable all native methods except `jni_thread_fifo`. Build the project and run it. The `logcat` output is shown in the following screenshot:

```
10-10 23:14:48.793 I/NativeThreadsSchedule(  742): process 742 thread 0 started, policy: 1, prio: 10
10-10 23:14:48.803 I/NativeThreadsSchedule(  755): process 755 thread 0 started, policy: 1, prio: 1
10-10 23:14:48.803 I/NativeThreadsSchedule(  755): process 755 thread 1 started, policy: 1, prio: 1
10-10 23:14:48.803 I/NativeThreadsSchedule(  755): process 755 thread 2 started, policy: 1, prio: 1
10-10 23:14:48.813 I/NativeThreadsSchedule(  755): process 755 thread 3 started, policy: 1, prio: 1
10-10 23:14:48.813 I/NativeThreadsSchedule(  755): process 755 thread 4 started, policy: 1, prio: 10
10-10 23:14:53.713 I/NativeThreadsSchedule(  755): process 755 thread 1: 6000000
10-10 23:14:53.773 I/NativeThreadsSchedule(  755): process 755 thread 3: 6000000
10-10 23:14:53.843 I/NativeThreadsSchedule(  742): process 742 thread 0: 6000000
10-10 23:14:53.863 I/NativeThreadsSchedule(  755): process 755 thread 4: 6000000
10-10 23:14:53.873 I/NativeThreadsSchedule(  755): process 755 thread 0: 6000000
10-10 23:14:53.952 I/NativeThreadsSchedule(  755): process 755 thread 2: 6000000
10-10 23:15:00.532 I/NativeThreadsSchedule(  755): process 755 thread 1: 12000000
10-10 23:15:00.603 I/NativeThreadsSchedule(  755): process 755 thread 3: 12000000
10-10 23:15:00.643 I/NativeThreadsSchedule(  755): process 755 thread 4: 12000000
10-10 23:15:00.693 I/NativeThreadsSchedule(  742): process 742 thread 0: 12000000
10-10 23:15:00.733 I/NativeThreadsSchedule(  755): process 755 thread 2: 12000000
10-10 23:15:00.763 I/NativeThreadsSchedule(  755): process 755 thread 0: 12000000
10-10 23:15:07.043 I/NativeThreadsSchedule(  755): process 755 thread 1 return
10-10 23:15:07.073 I/NativeThreadsSchedule(  755): process 755 thread 4 return
10-10 23:15:07.123 I/NativeThreadsSchedule(  755): process 755 thread 3 return
10-10 23:15:07.193 I/NativeThreadsSchedule(  742): process 742 thread 0 return
10-10 23:15:07.233 I/NativeThreadsSchedule(  755): process 755 thread 0 return
10-10 23:15:07.253 I/NativeThreadsSchedule(  755): process 755 thread 2 return
```

8. In `MainActivity.java`, disable all native methods except `jni_thread_set_priority`. Build the project and run it. The `logcat` output is shown in the following screenshot:

```
10-10 23:18:31.333 I/NativeThreadsSchedule(  795): thread 0 started
10-10 23:18:31.333 I/NativeThreadsSchedule(  795): thread 1 started
10-10 23:18:31.333 I/NativeThreadsSchedule(  795): thread 2 started
10-10 23:18:31.333 I/NativeThreadsSchedule(  795): thread 3 started
10-10 23:18:31.343 I/NativeThreadsSchedule(  795): thread 4 started
10-10 23:18:32.133 I/NativeThreadsSchedule(  795): thread 4 (prio = -20): 6000000
10-10 23:18:33.303 I/NativeThreadsSchedule(  795): thread 4 (prio = -20): 12000000
10-10 23:18:34.483 I/NativeThreadsSchedule(  795): thread 4 (prio = -20): 18000000 return
10-10 23:18:34.913 I/NativeThreadsSchedule(  795): thread 3 (prio = -10): 6000000
10-10 23:18:36.123 I/NativeThreadsSchedule(  795): thread 3 (prio = -10): 12000000
10-10 23:18:37.312 I/NativeThreadsSchedule(  795): thread 3 (prio = -10): 18000000 return
10-10 23:18:37.772 I/NativeThreadsSchedule(  795): thread 2 (prio = 0): 6000000
10-10 23:18:38.953 I/NativeThreadsSchedule(  795): thread 2 (prio = 0): 12000000
10-10 23:18:40.123 I/NativeThreadsSchedule(  795): thread 2 (prio = 0): 18000000 return
10-10 23:18:40.593 I/NativeThreadsSchedule(  795): thread 1 (prio = 10): 6000000
10-10 23:18:41.812 I/NativeThreadsSchedule(  795): thread 1 (prio = 10): 12000000
10-10 23:18:43.033 I/NativeThreadsSchedule(  795): thread 1 (prio = 10): 18000000 return
10-10 23:18:43.353 I/NativeThreadsSchedule(  795): thread 0 (prio = 19): 6000000
10-10 23:18:44.403 I/NativeThreadsSchedule(  795): thread 0 (prio = 19): 12000000
10-10 23:18:45.463 I/NativeThreadsSchedule(  795): thread 0 (prio = 19): 18000000 return
```

9. In `MainActivity.java`, disable all native methods except `jni_process_setThreadPriority`. Build the project and run it. The `logcat` output is shown in the following screenshot:

```
10-10 23:20:59.303 I/NativeThreadsSchedule(  846): thread 0 started
10-10 23:20:59.303 I/NativeThreadsSchedule(  846): thread 1 started
10-10 23:20:59.303 I/NativeThreadsSchedule(  846): thread 2 started
10-10 23:20:59.303 I/NativeThreadsSchedule(  846): thread 3 started
10-10 23:20:59.303 I/NativeThreadsSchedule(  846): thread 4 started
10-10 23:21:00.143 I/NativeThreadsSchedule(  846): thread 4 (prio = -20): 6000000
10-10 23:21:01.363 I/NativeThreadsSchedule(  846): thread 4 (prio = -20): 12000000
10-10 23:21:02.542 I/NativeThreadsSchedule(  846): thread 4 (prio = -20): 18000000 return
10-10 23:21:02.973 I/NativeThreadsSchedule(  846): thread 3 (prio = -10): 6000000
10-10 23:21:04.163 I/NativeThreadsSchedule(  846): thread 3 (prio = -10): 12000000
10-10 23:21:05.353 I/NativeThreadsSchedule(  846): thread 3 (prio = -10): 18000000 return
10-10 23:21:05.813 I/NativeThreadsSchedule(  846): thread 2 (prio = 0): 6000000
10-10 23:21:06.953 I/NativeThreadsSchedule(  846): thread 2 (prio = 0): 12000000
10-10 23:21:08.043 I/NativeThreadsSchedule(  846): thread 2 (prio = 0): 18000000 return
10-10 23:21:08.863 I/NativeThreadsSchedule(  846): thread 1 (prio = 10): 6000000
10-10 23:21:10.043 I/NativeThreadsSchedule(  846): thread 1 (prio = 10): 12000000
10-10 23:21:11.233 I/NativeThreadsSchedule(  846): thread 1 (prio = 10): 18000000 return
10-10 23:21:11.542 I/NativeThreadsSchedule(  846): thread 0 (prio = 19): 6000000
10-10 23:21:12.582 I/NativeThreadsSchedule(  846): thread 0 (prio = 19): 12000000
10-10 23:21:13.652 I/NativeThreadsSchedule(  846): thread 0 (prio = 19): 18000000 return
```

## How it works...

We can schedule native threads by setting the scheduling contention scope, thread priority, and scheduling policy:

- **Scheduling contention scope**: It determines the threads that a thread must compete against when the scheduler schedules threads

- **Thread priority**: It determines which thread is more likely to be selected by the scheduler when a CPU is available

- **Scheduling policy**: It determines how the scheduler schedules threads with the same priority

One way to adjust these settings is through the thread attribute. The following functions are defined in `pthread.h` to initialize and destroy an instance of `pthread_attr_t`:

```
int pthread_attr_init(pthread_attr_t * attr);
int pthread_attr_destroy(pthread_attr_t * attr);
```

In these two functions, the input argument is a pointer to a `pthread_attr_t` object. We will now describe contention scope, thread priority, and scheduling policy in detail.

## Scheduling contention scope

Two scopes are defined in a typical pthread implementation, namely `PTHREAD_SCOPE_SYSTEM` and `PTHREAD_SCOPE_PROCESS`. A system scope thread competes for the CPU with all other threads system-wide. On the other hand, a process scope thread is scheduled against other threads in the same process.

Android Bionic `pthread.h` defines the following two functions to set and get the thread scope:

```
int pthread_attr_setscope(pthread_attr_t *attr, int  scope);
int pthread_attr_getscope(pthread_attr_t const *attr);
```

The two functions accept a pointer to a pthread attribute object as the input argument. The `set` function also includes a second argument to let us pass the scope constant. These two functions return a zero to indicate success and a nonzero value to signal failure.

It turns out `pthread_attr_setscope` with `PTHREAD_SCOPE_PROCESS` as second input argument is not supported by Android. In other words, Android native threads always have system scope. As shown in `jni_thread_scope` at `NativeThreadsSchedule.cpp`, calling `pthread_attr_setscope` with `PTHREAD_SCOPE_PROCESS` will return a nonzero value.

We demonstrated the usage of the two functions previously in the native method `jni_thread_scope`. We created two processes in the method. The child process runs five threads, and the parent process only runs a single thread. Because they are all system scope threads, the threads are scheduled to get roughly same amount of CPU time slices regardless of the process they belong to, and therefore they finish at roughly the same time as shown in the first `logcat` screenshot in step 6 of the *How to do it...* section of this recipe.

>  We called `fork` to create a process in our code. This is for demonstration purpose. It is strongly discouraged to create a native process with `fork` on Android because the native process won't be managed by the Android framework and a misbehaving native process can consume lots of CPU cycles and cause security vulnerabilities.

## Scheduling policy and thread priority

Each thread has an associated scheduling policy and priority. A thread with higher priority is more likely to be selected by the scheduler when a CPU is available. In case multiple threads have the same priority, the scheduling policy will determine how to schedule them. The policies defined in Android `pthread.h` include `SCHED_OTHER`, `SCHED_FIFO`, and `SCHED_RR`.

The valid range of priority values is associated with the scheduling policy.

SCHED_FIFO: In the **First In First Out (FIFO)** policy, a thread gets the CPU until it exits or blocks. If blocked, it is placed at the end of the queue for its priority and the front thread in the queue will be given to the CPU. The priority range allowed under this policy is 1 to 99.

SCHED_RR: The **Round Robin (RR)** policy is similar to FIFO except that each thread is only allowed to run for a certain amount of time, known as quantum. When a thread finishes its quantum, it is interrupted and placed at the end of the queue for its priority. The priority range allowed under this policy is also 1 to 99.

SCHED_OTHER: This is the default scheduling policy. It also allows a thread to run only a limited amount of times, but the algorithm can be different and more complicated than SCHED_RR. All threads under this policy have a priority of 0.

People who are experienced with pthreads programming may be familiar with the pthreads policy and priority functions including:

- pthread_attr_setschedpolicy
- pthread_attr_getschedpolicy
- pthread_attr_setschedparam
- pthread_attr_getschedparam

These functions do not work on Android, as expected, although they are defined in the Android pthread.h header. Therefore, we will not discuss the details here.

In our sample project, we implemented a native method jni_thread_fifo, which attempts to set the scheduling policy as SCHED_FIFO and the thread priority. As shown in the second logcat screenshot, the threads are not affected by these settings.

In summary, all Android threads are system scope threads with 0 priority, and scheduling policy SCHED_OTHER.

## Scheduling using nice value/level

Nice value/level is another factor that can affect the scheduler. It is also often referred to as priority, but here we will use nice value to differentiate it with the thread priority we discussed earlier.

We use the following two approaches to adjust the nice value:

▶ **Calling setpriority**: This is demonstrated in `SetPriority.cpp`. We created five threads with different nice values, and the third `logcat` screenshot in step 8 of the *How to do it* section indicates the thread with lower nice values return first.

▶ **Calling android.os.Process.setThreadPriority**: This is illustrated in `JNIProcessSetThreadPriority.cpp`. As shown in the fourth `logcat` screenshot at step 9 of the *How to do it* section, we got similar result as calling `setpriority`. In fact, `setThreadPriority` calls `setpriority` internally.

# Managing data for native threads at Android NDK

There are several options when we want to preserve thread-wide data across functions, including global variables, argument passing, and thread-specific data key. This recipe discusses all the three options with a focus on thread-specific data key.

## Getting ready...

Readers are recommended to read the *Creating and terminating native threads at Android NDK* recipe and the *Synchronizing native threads with mutex at Android NDK* recipe in this chapter before this one.

## How to do it...

The following steps will help us create an Android project that demonstrates data management at Android NDK:

1. Create an Android application named `NativeThreadsData`. Set the package name as `cookbook.chapter6.nativethreadsdata`. Please refer to the *Loading native libraries and registering native methods* recipe in *Chapter 2, Java Native Interface* if you want more detailed instructions.

2. Right-click on the project **NativeThreadsData**, select **Android Tools | Add Native Support**.

3. Add a Java file named `MainActivity.java` under package `cookbook.chapter6.nativethreadsdata`. This Java file simply loads the native library `NativeThreadsData` and calls the native methods.

4. Add `mylog.h` and `NativeThreadsData.cpp` files under the `jni` folder. The `mylog.h` contains the Android native `logcat` utility functions, while the `NativeThreadsData.cpp` file contains the native code to start multiple threads. A part of the code is shown as follows:

`jni_start_threads` starts *n* number of threads, where *n* is specified by the variable `pNumOfThreads`:

```
void jni_start_threads(JNIEnv *pEnv, jobject pObj, int
pNumOfThreads) {
  pthread_t *ths;
  int i, ret;
  int *thNum;
  ths = (pthread_t*)malloc(sizeof(pthread_t)*pNumOfThreads);
  thNum = (int*)malloc(sizeof(int)*pNumOfThreads);
  pthread_mutex_init(&mux, NULL);
  pthread_key_create(&muxCntKey, free_muxCnt);
  for (i = 0; i < pNumOfThreads; ++i) {
    thNum[i] = i;
    ret = pthread_create(&ths[i], NULL, run_by_thread,
(void*)&(thNum[i]));
  }
  for (i = 0; i < pNumOfThreads; ++i) {
    ret = pthread_join(ths[i], NULL);
  }
  pthread_key_delete(muxCntKey);
  pthread_mutex_destroy(&mux);
  free(thNum);
  free(ths);
}
```

The `thread_step_1` function is executed by threads. It gets the data associated with the thread-specific key and uses it to count the number of times the mutex is locked:

```
void thread_step_1() {
  struct timeval st, cu;
  long stt, cut;
  int *muxCntData = (int*)pthread_getspecific(muxCntKey);
  gettimeofday(&st, NULL);
  stt = st.tv_sec*1000 + st.tv_usec/1000;
  do {
          pthread_mutex_lock(&mux);
    (*muxCntData)++;
          pthread_mutex_unlock(&mux);
    gettimeofday(&st, NULL);
    cut = st.tv_sec*1000 + st.tv_usec/1000;
    } while (cut - stt < 10000);
}
```

The `thread_step_2` function is executed by threads. It gets the data associated with the thread-specific key and prints it out:

```
void thread_step_2(int thNum) {
    int *muxCntData = (int*)pthread_getspecific(muxCntKey);
    LOGI(1, "thread %d: mux usage count: %d\n", thNum, *muxCntData);
}
```

The `run_by_thread` function is executed by threads:

```
void *run_by_thread(void *arg) {
    int* threadNum = (int*)arg;
    int *muxCntData = (int*)malloc(sizeof(int));
    *muxCntData = 0;
    pthread_setspecific(muxCntKey, (void*)muxCntData);
    thread_step_1();
    thread_step_2(*threadNum);
    return NULL;
}
```

5. Add an `Android.mk` file under the `jni` folder with the following content:

```
LOCAL_PATH := $(call my-dir)
include $(CLEAR_VARS)
LOCAL_MODULE    := NativeThreadsData
LOCAL_SRC_FILES := NativeThreadsData.cpp
LOCAL_LDLIBS    := -llog
include $(BUILD_SHARED_LIBRARY)
```

6. Build and run the Android project, and use the following command to monitor the `logcat` output:

```
$ adb logcat -v time NativeThreadsData:I *:S
```

The `logcat` output is shown in the following screenshot:

```
10-11 22:54:36.376 I/NativeThreadsData(21881): thread 0 started
10-11 22:54:36.376 I/NativeThreadsData(21881): thread 1 started
10-11 22:54:36.376 I/NativeThreadsData(21881): thread 2 started
10-11 22:54:46.386 I/NativeThreadsData(21881): thread 1: mux usage count: 3037905
10-11 22:54:46.386 I/NativeThreadsData(21881): thread 2: mux usage count: 3021446
10-11 22:54:46.386 I/NativeThreadsData(21881): 2 return
10-11 22:54:46.386 I/NativeThreadsData(21881): thread 0: mux usage count: 3015422
10-11 22:54:46.386 I/NativeThreadsData(21881): 0 return
10-11 22:54:46.386 I/NativeThreadsData(21881): 1 return
```

## How it works...

In our sample project, we demonstrated passing data using global variables, argument, and thread-specific data key:

- The mutex `mux` is declared as a global variable, and each thread can access it.
- Each thread is assigned a thread number as input argument. In the function `run_by_thread`, each thread passes the accepted thread number to another function `thread_step_2`.
- We defined a thread-specific key `muxCntKey`. Each thread can associate its own value with the key. In our code, we used this key to store the number of times a thread locks the mutex `mux`.

Next we'll discuss the thread-specific data key in detail.

### Creation and deletion of thread-specific data key

The following two functions are defined in `pthread.h` to create and delete a thread-specific data key respectively:

```
int pthread_key_create(pthread_key_t *key, void (*destructor_function)
(void *));
int pthread_key_delete (pthread_key_t key);
```

`pthread_key_create` accepts a pointer to the `pthread_key_t` structure and a function pointer to a destruction function to be associated with each key value. The destruction function is optional and can be set to `NULL`. In our example, we called `pthread_key_create` to create the key named `muxCntKey`.

The `pthread_key_create` function returns zero to indicate success and some other values for failure. If successful, the first input argument `key` will be pointing to the newly created key, and the value `NULL` is associated with the new key in all active threads. If a new thread is created after the key creation, the value `NULL` is also associated with the key for the new thread.

When a thread exits, the associated value of the key is set to `NULL`, and then the destruction function associated with the key is called with the key's previously associated value as the sole input argument. In our sample code, we defined a destruction function `free_muxCnt` to free the memory of data associated with the key `muxCntKey`.

`pthread_key_delete` is relatively simple. It accepts a key created by `pthread_key_create` and deletes it. It returns zero for success and a nonzero value for failure.

## Set and get thread-specific data

Android `pthread.h` defines the following two functions for thread-specific data management:

```
int pthread_setspecific(pthread_key_t key, const void *value);
void *pthread_getspecific(pthread_key_t key);
```

The `pthread_setspecific` function accepts a previously created data key and a pointer to data to be associated with the key. It returns a zero to indicate success and nonzero otherwise. Different threads can call this function to bind different values to the same key.

`pthread_getspecific` accepts a previously created data and key and returns a pointer to the data associated with the key in the calling thread.

In the `run_by_thread` function of our sample code, we associate an integer variable initialized to zero to the `muxCntKey` key. In function `thread_step_1`, we get the integer variable associated with the key and use it to count the number of times `mux` is locked. In function `thread_step_2`, we again obtain the integer variable associated with `muxCntKey` and print its value.

# 7
# Other Android NDK API

In this chapter we will cover:

- ▶ Programming with the jnigraphics library in Android NDK
- ▶ Programming with the dynamic linker library in Android NDK
- ▶ Programming with the zlib compression library in Android NDK
- ▶ Programming audio with the OpenSL ES audio library in Android NDK
- ▶ Programming with the OpenMAX AL multimedia library in Android NDK

## Introduction

In the previous three chapters, we have covered Android NDK OpenGL ES API (*Chapter 4, Android NDK OpenGL ES API*), Native Application API (*Chapter 5, Android Native Application API*), and Multithreading API (*Chapter 6, Android NDK Multithreading*). This is the last chapter on Android NDK API illustration, and we will cover a few more libraries, including the jnigraphics library, dynamic linker library, zlib compression library, OpenSL ES Audio library, and OpenMAX AL multimedia library.

We first introduce two small libraries, jnigraphics and dynamic linker, which only have a few API functions and are easy to use. We then describe zlib compression library, which can be used to compress and decompress data in .zlib and .gzip formats. The OpenSL ES audio library and OpenMAX AL multimedia library are two relatively new APIs available on newer versions of Android. The API functions in these two libraries are not frozen yet and still evolving. Because the source compatibility is not a goal of the library development on Android, as stated in the NDK OpenSL ES and OpenMAX AL documentation, future versions of these two libraries may require us to update our code.

Also, note that OpenSL ES and OpenMAX AL are fairly complex libraries with lots of API functions. We can only introduce the basic usage of these two libraries with simple examples. Interested readers should refer to the library documentation for more details.

# Programming with the jnigraphics library in Android NDK

The `jnigraphics` library provides a C-based interface for native code to access the pixel buffers of Java bitmap objects, which is available as a stable native API on Android 2.2 system images and higher. This recipe discusses how to use the `jnigraphics` library.

## Getting ready...

Readers are expected to know how to create an Android NDK project. We can refer to the *Writing a Hello NDK program* recipe of *Chapter 1, Hello NDK* for detailed instructions.

## How to do it...

The following steps describe how to create a simple Android application which demonstrates the usage of the `jnigraphics` library:

1. Create an Android application named `JNIGraphics`. Set the package name as `cookbook.chapter7.JNIGraphics`. Refer to the *Loading native libraries and registering native methods* recipe of *Chapter 2, Java Native Interface* for more detailed instructions.

2. Right-click on the project **JNIGraphics**, select **Android Tools | Add Native Support**.

3. Add two Java files named `MainActivity.java` and `RenderView.java` in the `cookbook.chapter7.JNIGraphics` package. The `RenderView.java` loads the `JNIGraphics` native library, calls the native `naDemoJniGraphics` method to process a bitmap, and finally display the bitmap. The `MainActivity.java` files creates a bitmap, passes it to the `RenderView` class, and sets the `RenderView` class as its content view.

4. Add `mylog.h` and `JNIGraphics.cpp` files under the `jni` folder. The `mylog.h` contains the Android native `logcat` utility functions, while the `JNIGraphics.cpp` file contains the native code to process the bitmap with the `jnigraphics` library functions. A part of the code in the `JNIGraphics.cpp` file is shown as follows:

```
void naDemoJniGraphics(JNIEnv* pEnv, jclass clazz, jobject
pBitmap) {
    int lRet, i, j;
    AndroidBitmapInfo lInfo;
    void* lBitmap;
    //1. retrieve information about the bitmap
```

```
   if ((lRet = AndroidBitmap_getInfo(pEnv, pBitmap, &lInfo)) < 0) {
      return;
   }
   if (lInfo.format != ANDROID_BITMAP_FORMAT_RGBA_8888) {
      return;
   }
   //2. lock the pixel buffer and retrieve a pointer to it
   if ((lRet = AndroidBitmap_lockPixels(pEnv, pBitmap, &lBitmap)) <
0) {
      LOGE(1, "AndroidBitmap_lockPixels() failed! error = %d",
lRet);
   }
   //3. manipulate the pixel buffer
   unsigned char *pixelBuf = (unsigned char*)lBitmap;
   for (i = 0; i < lInfo.height; ++i) {
      for (j = 0; j < lInfo.width; ++j) {
      unsigned char *pixelP = pixelBuf + i*lInfo.stride + j*4;
      *pixelP = (unsigned char)0x00;        //remove R component
//    *(pixelP+1) = (unsigned char)0x00;        //remove G component
//    *(pixelP+2) = (unsigned char)0x00;        //remove B component
//    LOGI(1, "%d:%d:%d:%d", *pixelP, *(pixelP+1), *(pixelP+2),
*(pixelP+3));
      }
   }
   //4. unlock the bitmap
   AndroidBitmap_unlockPixels(pEnv, pBitmap);
}
```

5. Add an `Android.mk` file in the `jni` folder with the following content:

```
LOCAL_PATH := $(call my-dir)
include $(CLEAR_VARS)
LOCAL_MODULE      := JNIGraphics
LOCAL_SRC_FILES := JNIGraphics.cpp
LOCAL_LDLIBS := -llog -ljnigraphics
include $(BUILD_SHARED_LIBRARY)
```

6. Build and run the Android project. We can enable code to remove different components from the bitmap. The following screenshots show the original picture and the ones with red, green, and blue component removed respectively:

## How it works...

In our sample project, we modified the bitmap passed to the native `naDemoJniGraphics` function by setting one of its RGB components to zero.

 The `jnigraphics` library is only available for Android API level 8 (Android 2.2, Froyo) and higher.

The following steps should be followed to use the `jnigraphics` library:

1. Include the `<android/bitmap.h>` header in the source code where we use the `jnigraphics` API.

2. Link to the `jnigraphics` library by including the following line in the `Android.mk` file.

   ```
   LOCAL_LDLIBS += -ljnigraphics
   ```

3. In the source code, call `AndroidBitmap_getInfo` to retrieve the information about a bitmap object. The `AndroidBitmap_getInfo` function has the following prototype:

   ```
   int AndroidBitmap_getInfo(JNIEnv* env, jobject jbitmap,
   AndroidBitmapInfo* info);
   ```

   The function accepts a pointer to the `JNIEnv` structure, a reference to the bitmap object, and a pointer to the `AndroidBitmapInfo` structure. If the call is successful, the data structure pointed by `info` will be filled.

The AndroidBitmapInfo is defined as follows:

```
typedef struct {
uint32_t    width;
     uint32_t    height;
uint32_t    stride;
int32_t     format;
uint32_t    flags;
} AndroidBitmapInfo;
```

width and height indicate the pixel width and height of the bitmap. stride refers to the number of bytes to skip between rows of the pixel buffer. The number must be no less than the width in bytes. In most cases, stride is the same as width. However, sometimes pixel buffer contains paddings so stride can be bigger than bitmap width.

The format is the color format, which can be ANDROID_BITMAP_FORMAT_RGBA_8888, ANDROID_BITMAP_FORMAT_RGB_565, ANDROID_BITMAP_FORMAT_RGBA_4444, ANDROID_BITMAP_FORMAT_A_8, or ANDROID_BITMAP_FORMAT_NONE as defined in the bitmap.h header file.

In our example, we used ANDROID_BITMAP_FORMAT_RGBA_8888 as the bitmap format. Therefore, every pixel takes 4 bytes.

4.  Lock the pixel address by calling the AndroidBitmap_lockPixels function:

    ```
    int AndroidBitmap_lockPixels(JNIEnv* env, jobject jbitmap, void** addrPtr);
    ```

    If the call succeeds, the *addrPtr pointer will point to the pixels of the bitmap. Once the pixel address is locked, the memory for the pixels will not move until the pixel address is unlocked.

5.  Manipulate the pixel buffer in the native code.

6.  Unlock the pixel address by calling AndroidBitmap_unlockPixels:

    ```
    int AndroidBitmap_unlockPixels(JNIEnv* env, jobject jbitmap);
    ```

    Note that this function must be called if the AndroidBitmap_lockPixels function succeeds.

 The jnigraphics functions return ANDROID_BITMAP_RESUT_SUCCESS, which has a value of 0, upon success. They return a negative value upon failure.

## There's more...

Recall that we used the `jnigraphics` library to load textures in the *Mapping texture to 3D objects with OpenGL ES 1.x API* recipe in *Chapter 4, Android NDK OpenGL ES API*. We can revisit the recipe for another example of how we use the `jnigraphics` library.

# Programming with the dynamic linker library in Android NDK

Dynamic loading is a technique to load a library into memory at runtime, and execute functions or access variables defined in the library. It allows the app to start without these libraries.

We have seen dynamic loading in almost every recipe of this book. When we call the `System.loadLibrary` or `System.load` function to load the native libraries, we are using dynamic loading.

Android NDK has provided the dynamic linker library to support dynamic loading in NDK, since Android 1.5. This recipe discusses the dynamic linker library functions.

## Getting ready...

Readers are expected to know how to create an Android NDK project. You can refer to the *Writing a Hello NDK program* recipe of *Chapter 1, Hello NDK* for detailed instructions.

## How to do it...

The following steps describe how to create an Android application using the dynamic linking library to load the math library and compute the square root of 2.

1. Create an Android application named `DynamicLinker`. Set the package name as `cookbook.chapter7.dynamiclinker`. Refer to the *Loading native libraries and registering native methods* recipe of *Chapter 2, Java Native Interface* for more detailed instructions.

2. Right-click on the `DynamicLinker` project, select **Android Tools | Add Native Support**.

3. Add a Java file named `MainActivity.java` under the `cookbook.chapter7.dynamiclinker` package. This Java file simply loads the native `DynamicLinker` library and calls the native `naDLDemo` method.

4. Add the `mylog.h` and `DynamicLinker.cpp` files under the `jni` folder. A part of the code in the `OpenSLESDemo.cpp` file is shown in the following code.

   `naDLDemo` loads the `libm.so` library, obtains the address of the `sqrt` function and calls the function with input argument `2.0`:

```
void naDLDemo(JNIEnv* pEnv, jclass clazz) {
  void *handle;
  double (*sqrt)(double);
  const char *error;
  handle = dlopen("libm.so", RTLD_LAZY);
  if (!handle) {
    LOGI(1, "%s\n", dlerror());
    return;
  }
  dlerror();    /* Clear any existing error */
  *(void **) (&sqrt) = dlsym(handle, "sqrt");
  if ((error = dlerror()) != NULL)  {
    LOGI(1, "%s\n", error);
    return;
  }
  LOGI(1, "%f\n", (*sqrt)(2.0));
}
```

5. Add an `Android.mk` file under the `jni` folder with the following content:

```
LOCAL_PATH := $(call my-dir)
include $(CLEAR_VARS)
LOCAL_MODULE     := DynamicLinker
LOCAL_SRC_FILES := DynamicLinker.cpp
LOCAL_LDLIBS := -llog -ldl
include $(BUILD_SHARED_LIBRARY)
```

6. Build and run the Android project, and use the following command to monitor the `logcat` output:

```
$ adb logcat -v time DynamicLinker:I *:S
```

A screenshot of the `logcat` output is shown as follows:

```
10-26 22:56:39.724 I/DynamicLinker( 2892): 1.414214
```

## How it works...

In order to build with dynamic loading library `libdl.so`, we must add the following line to the `Android.mk` file:

```
LOCAL_LDLIBS := -ldl
```

The following functions are defined in the `dlfcn.h` header file by the Android dynamic linking library:

```
void*       dlopen(const char*  filename, int flag);
int         dlclose(void*  handle);
const char* dlerror(void);
void*       dlsym(void*  handle, const char*  symbol);
int         dladdr(const void* addr, Dl_info *info);
```

The `dlopen` function loads the library dynamically. The first argument indicates the library name, while the second argument refers to the loading mode, which describes how `dlopen` resolves the undefined symbols. When an object file (for example, shared library, executable file, and so on) is loaded, it may contain references to symbols whose addresses are not known until another object file is loaded (such symbols are referred to as undefined symbols). These references need to be resolved before they can be used to access the symbols. The following two modes determine when the resolving happens:

> ▸ RTLD_NOW: When the object file is loaded, the undefined symbols are resolved. This means the resolving occurs before the `dlopen` function returns. This may be a waste if resolving is performed but the references are never accessed.

> ▸ RTLD_LAZY: The resolving can be performed after the `dlopen` function returns, that is, the undefined symbols are resolved when the code is executed.

The following two modes determine the visibility of the symbols in the loaded object. They can be ORed with the previously mentioned two modes:

> ▸ RTLD_LOCAL: The symbols will not be available for another object

> ▸ RTLD_GLOBAL: The symbols will be available for subsequently loaded objects

The `dlopen` function returns a handle upon success. The handle should be used for the subsequent calls to `dlsym` and `dlclose`.

The `dlclose` function simply decrements the reference count of the loaded library handle. If the reference count is reduced to zero, the library will be unloaded.

The `dlerror` function returns a string to describe the most recent error occurred while calling `dlopen`, `dlsym`, or `dlclose` since the last call to `dlerror`. It returns NULL if no such error occurred.

The `dlsym` function returns the memory address of a given symbol of the loaded dynamic library referred by the input argument handle. The returned address can be used to access the symbol.

The `dladdr` function takes an address and tries to return more information about the address and library through the `info` argument of the `DI_info` type. The `DI_info` data structure is defined as shown in the following code snippet:

```
typedef struct {
    const char *dli_fname;
    void        *dli_fbase;
    const char *dli_sname;
    void        *dli_saddr;
} Dl_info;
```

`dli_fname` indicates the path of the shared object referred by the input argument `addr`. The `dli_fbase` is the address where the shared object is loaded. `dli_sname` indicates the name of the nearest symbol with address lower than `addr`, and `dli_saddr` is the address of symbol named by `dli_sname`.

In our example, we demonstrated the usage of the first four functions. We load the math library by `dlopen`, obtain the address of the `sqrt` function by `dlsym`, check the error by `dlerror`, and close the library by `dlclose`.

For more details on the dynamic loading library, refer to `http://tldp.org/HOWTO/Program-Library-HOWTO/dl-libraries.html` and `http://linux.die.net/man/3/dlopen`.

# Programming with the zlib compression library in Android NDK

`zlib` is a widely-used, lossless data compression library, which is available for Android 1.5 system images or higher. This recipe discusses the basic usage of the `zlib` functions.

## Getting ready...

Readers are expected to know how to create an Android NDK project. We can refer to the *Writing a Hello NDK program* recipe of *Chapter 1, Hello NDK* for detailed instructions.

## How to do it...

The following steps describe how to create a simple Android application which demonstrates the usage of `zlib` library:

1. Create an Android application named `ZlibDemo`. Set the package name as `cookbook.chapter7.zlibdemo`. Refer to the *Loading native libraries and registering native methods* recipe of *Chapter 2, Java Native Interface* for more detailed instructions.

2. Right-click on the project **ZlibDemo**, select **Android Tools | Add Native Support**.

3. Add a Java file named `MainActivity.java` in the `cookbook.chapter7.zlibdemo` package. The `MainActivity.java` file loads the `ZlibDemo` native library, and calls the native methods.

4. Add `mylog.h`, `ZlibDemo.cpp`, and `GzFileDemo.cpp` files under the `jni` folder. The `mylog.h` header file contains the Android native `logcat` utility functions, while `ZlibDemo.cpp` and `GzFileDemo.cpp` files contain code for compression and decompression. A part of the code in `ZlibDemo.cpp` and `GzFileDemo.cpp` is shown in the following code.

`ZlibDemo.cpp` contains the native code to compress and decompress data in memory.

`compressUtil` compresses and decompress data in memory.

```
void compressUtil(unsigned long originalDataLen) {
    int rv;
    int compressBufBound = compressBound(originalDataLen);
    compressedBuf = (unsigned char*) malloc(sizeof(unsigned
char)*compressBufBound);
    unsigned long compressedDataLen = compressBufBound;
    rv = compress2(compressedBuf, &compressedDataLen, dataBuf,
originalDataLen, 6);
    if (Z_OK != rv) {
        LOGE(1, "compression error");
        free(compressedBuf);
        return;
    }
    unsigned long decompressedDataLen = S_BUF_SIZE;
    rv = uncompress(decompressedBuf, &decompressedDataLen,
compressedBuf, compressedDataLen);
    if (Z_OK != rv) {
        LOGE(1, "decompression error");
        free(compressedBuf);
        return;
    }
```

```
  if (0 == memcmp(dataBuf, decompressedBuf, originalDataLen)) {
    LOGI(1, "decompressed data same as original data");
  }    //free resource
  free(compressedBuf);
}
```

5. `naCompressAndDecompress` generates data for compression and calls the `compressUtil` function to compress and decompress the generated data:

```
void naCompressAndDecompress(JNIEnv* pEnv, jclass clazz) {
  unsigned long originalDataLen = getOriginalDataLen();
  LOGI(1, "---------data with repeated bytes---------")
  generateOriginalData(originalDataLen);
  compressUtil(originalDataLen);
  LOGI(1, "---------data with random bytes---------")
  generateOriginalDataRandom(originalDataLen);
  compressUtil(originalDataLen);
}
```

`GzFileDemo.cpp` contains the native code to compress and decompress the data in file.

`writeToFile` writes a string to a `gzip` file. Compression is applied at writing:

```
int writeToFile() {
  gzFile file;
  file = gzopen("/sdcard/test.gz", "w6");
  if (NULL == file) {
    LOGE(1, "cannot open file to write");
    return 0;
  }
  const char* dataStr = "hello, Android NDK!";
  int bytesWritten = gzwrite(file, dataStr, strlen(dataStr));
  gzclose(file);
  return bytesWritten;
}
```

`readFromFile` reads data from the `gzip` file. Decompression is applied at reading:

```
void readFromFile(int pBytesToRead) {
  gzFile file;
  file = gzopen("/sdcard/test.gz", "r6");
  if (NULL == file) {
    LOGE(1, "cannot open file to read");
    return;
  }
```

```
    char readStr[100];
    int bytesRead = gzread(file, readStr, pBytesToRead);
    gzclose(file);
    LOGI(1, "%d: %s", bytesRead, readStr);
}
```

6. Add an `Android.mk` file under the `jni` folder with the following content:

```
LOCAL_PATH := $(call my-dir)
include $(CLEAR_VARS)
LOCAL_MODULE    := ZlibDemo
LOCAL_SRC_FILES := ZlibDemo.cpp GzFileDemo.cpp
LOCAL_LDLIBS := -llog -lz
include $(BUILD_SHARED_LIBRARY)
```

7. Enable the `naCompressAndDecompress` function and disable the `naGzFileDemo` function, build and run the application. We can monitor the `logcat` output with the following command:

```
$ adb logcat -v time ZlibDemo:I *:S
```

The `logcat` output screenshot is shown as follows:

```
10-20 16:23:19.542 I/ZlibDemo(20788): ---------data with repeated bytes---------
10-20 16:23:20.063 I/ZlibDemo(20788): upper bound:8719092; input: 8716419; compressed: 8495
10-20 16:23:20.113 I/ZlibDemo(20788): decompressed: 8716419
10-20 16:23:20.143 I/ZlibDemo(20788): decompressed data same as original data
10-20 16:23:20.143 I/ZlibDemo(20788): ---------data with random bytes---------
10-20 16:23:23.226 I/ZlibDemo(20788): upper bound:8719092; input: 8716419; compressed: 8719085
10-20 16:23:23.266 I/ZlibDemo(20788): decompressed: 8716419
10-20 16:23:23.296 I/ZlibDemo(20788): decompressed data same as original data
```

Enable the `naGzFileDemo` function and disable the `naCompressAndDecompress` function, build and run the application. The `logcat` output is shown in the following screenshot:

```
10-20 17:01:26.013 I/ZlibDemo(21448): 19: hello, Android NDK!
```

## How it works...

The `zlib` library provides compression and decompression functions for both in-memory data and files. We demonstrated both use cases. In the `ZlibDemo.cpp` file, we created two data buffers, one with repeated bytes, and the other with random bytes. We compress and decompress the data with the following steps:

1. Compute the upper bound on the compressed size. This is done by the following function:

   ```
   uLong compressBound(uLong sourceLen);
   ```

   The function returns the maximum size of the compressed data after calling the `compress` or `compress2` function on `sourceLen` bytes of source data.

2. Allocate the memory for storing the compressed data.

3. Compress the data. This is done by the following function:

   ```
   int compress2(Bytef *dest,    uLongf *destLen, const Bytef *source,
   uLong sourceLen, int level);
   ```

   This function accepts five input arguments. `source` and `sourceLen` refer to the source data buffer and source data length. `dest` and `destLen` indicate the data buffer for storing the compressed data and size of this buffer. The value of `destLen` must be at least the value returned by `compressBound` when the function is called. When the function is returned, `destLen` is set to the actual size of the compressed data. The last input argument `level` can be a value between 0 and 9, where 1 gives best speed and 9 gives best compression. In our example, we set the value as 6 to compromise between speed and compression.

    We can also use the compress function to compress the data, which does not have the level input argument. Instead, it assumes a default level, which is equivalent to 6.

4. Decompress the data. This is done by using the `uncompress` function:

   ```
   int uncompress(Bytef *dest,    uLongf *destLen, const Bytef
   *source, uLong sourceLen);
   ```

   The input arguments have the same meaning as the `compress2` function.

5. Compare the decompressed data with the original data. This is just a simple check.

   By default, these functions use the `zlib` format for the compressed data.

   This library also supports reading and writing files in the `gzip` format. This is demonstrated in `GzFileDemo.cpp`. The usage of these functions is similar to the `stdio` functions for file reading and writing.

The steps we followed to write compressed data to a `gzip` file and then read the uncompressed data from it are shown as follows:

1.  Open a `gzip` file for writing. This is done by the following function:

    ```
    gzFile gzopen(const char *path, const char *mode);
    ```

    The function accepts a filename and open mode, and returns a `gzFile` object on success. The mode is similar to the `fopen` function, but with an optional compression level. In our example, we called the `gzopen` with `w6` to specify the compression level as 6.

2.  Write data to a `gzip` file. This is done by the following function:

    ```
    int gzwrite(gzFile file, voidpc buf, unsigned len);
    ```

    This function writes uncompressed data into the compressed file. The input argument `file` refers to the compressed file, `buf` refers to the uncompressed data buffer, and `len` indicates the number of bytes to write. The function returns the actual number of uncompressed data written.

3.  Close the `gzip` file. This is done by the following function:

    ```
    int ZEXPORT    gzclose(gzFile file);
    ```

    Calling this function will flush all pending output and close the compressed file.

4.  Open the file for reading. We passed `r6` to the `gzopen` function.

5.  Read data from the compressed file. This is done by the `gzread` function.

    ```
    int gzread(gzFile file, voidp buf, unsigned len);
    ```

    The function reads `len` number of bytes from file into `buf`. It returns the actual number of bytes read.

 The `zlib` library supports two compression formats, `zlib` and `gzip`. `zlib` is designed to be compact and fast, so it is best for use in memory and on communication channels. On the other hand, `gzip` is designed for single file compression on a filesystem, which has a larger header for maintaining the directory information, and uses a slower check method than `zlib`.

In order to use the `zlib` library, we must include the `zlib.h` header file in our source code and add the following line to `Android.mk` to link to the `libz.so` library:

```
LOCAL_LDLIBS := -lz
```

## There's more...

Recall in the *Managing assets at Android NDK* recipe in *Chapter 5, Android Native Application AP* , we compiled the `libpng` library, which requires the `zlib` library.

We only covered a few functions provided by the `zlib` library. For more information, you can refer to the `zlib.h` and `zconf.h` header files in the `platforms/android-<version>/arch-arm/usr/include/` folder. Detailed documentation for the `zlib` library can be found at `http://www.zlib.net/manual.html`.

# Programming audio with the OpenSL ES audio library in Android NDK

OpenSL ES is an application level audio library in C. Android NDK native audio APIs are based on the OpenSL ES 1.0.1 standard with Android specific extensions. The API is available for Android 2.3 or higher and some features are only supported on Android 4.0 or higher. The API functions in this libraries are not frozen yet and are still evolving. Future versions of this library may require us to update our code. This recipe introduces OpenSL ES APIs in the context of Android.

## Getting ready...

Before we start coding with OpenSL ES, it is essential to understand some basics of the library. OpenSL ES stands for **Open Sound Library** for embedded systems, which is a cross-platform, royalty-free, C language application level API for developers to access audio functionalities of embedded systems. The library specification defines features like audio playback and recording, audio effects and controls, 2D and 3D audio, advanced MIDI, and so on. Based on the features supported, OpenSL ES defines three profiles, including phone, music, and game.

However, the Android native audio API does not conform to any of the three profiles, because it does not implement all features from any of the profiles. In addition, Android implements some features specific to Android, such as the Android buffer queue. For a detailed description of what is supported on Android, we can refer to the OpenSL ES for Android documentation available with Android NDK under the `docs/opensles/` folder.

Although OpenSL ES API is implemented in C, it adopts an object-oriented approach by building the library based on objects and interfaces:

- **Object**: An object is an abstraction of a set of resources and their states. Every object has a type assigned at its creation, and the type determines the set of tasks the object can perform. It is similar to the class concept in C++.

- **Interface**: An interface is an abstraction of a set of features an object can provide. These features are exposed to us as a set of methods and the type of each interface determines the exact set of features exposed. In the code, the type of an interface is identified by the interface ID.

It is important to note that an object does not have actual representation in code. We change the object's states and access its features through interfaces. An object can have one or more interface instances. However, no two instances of a single object can be the same type. In addition, a given interface instance can only belong to one object. This relationship can be illustrated as shown in the following diagram:

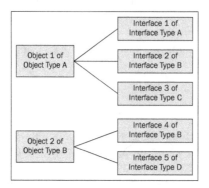

As shown in the diagram, Object 1 and Object 2 have different types and therefore expose different interfaces. Object 1 has three interface instances, all with different types. While Object 2 has another two interface instances with different types. Note that Interface 2 of Object 1 and Interface 4 of Object 2 have the same type, which means both Object 1 and Object 2 support the features exposed through interfaces of Interface Type B.

## How to do it...

The following steps describe how to create a simple Android application using the native audio library to record and play audio:

1. Create an Android application named OpenSLESDemo. Set the package name as cookbook.chapter7.opensles. Refer to the *Loading native libraries and registering native methods* recipe of *Chapter 2, Java Native Interface* for more detailed instructions.

2. Right-click on the project **OpenSLESDemo**, select **Android Tools | Add Native Support**.

3. Add a Java file named `MainActivity.java` in the `cookbook.chapter7.opensles` package. This Java file simply loads the native library `OpenSLESDemo` and calls the native methods to record and play audio.

4. Add `mylog.h`, `common.h`, `play.c`, `record.c`, and `OpenSLESDemo.cpp` files in the `jni` folder. A part of the code in the `play.c`, `record.c`, and `OpenSLESDemo.cpp` files is shown in the following code snippet.

`record.c` contains the code to create an audio recorder object and record the audio.

`createAudioRecorder` creates and realizes an audio player object and obtains the record and buffer queue interfaces:

```
jboolean createAudioRecorder() {
    SLresult result;
    SLDataLocator_IODevice loc_dev = {SL_DATALOCATOR_IODEVICE, SL_
IODEVICE_AUDIOINPUT, SL_DEFAULTDEVICEID_AUDIOINPUT, NULL};
    SLDataSource audioSrc = {&loc_dev, NULL};
    SLDataLocator_AndroidSimpleBufferQueue loc_bq = {SL_
DATALOCATOR_ANDROIDSIMPLEBUFFERQUEUE, 1};
    SLDataFormat_PCM format_pcm = {SL_DATAFORMAT_PCM, 1, SL_
SAMPLINGRATE_16,
        SL_PCMSAMPLEFORMAT_FIXED_16, SL_PCMSAMPLEFORMAT_FIXED_16,
        SL_SPEAKER_FRONT_CENTER, SL_BYTEORDER_LITTLEENDIAN};
    SLDataSink audioSnk = {&loc_bq, &format_pcm};
    const SLInterfaceID id[1] = {SL_IID_ANDROIDSIMPLEBUFFERQUEUE};
    const SLboolean req[1] = {SL_BOOLEAN_TRUE};
    result = (*engineEngine)->CreateAudioRecorder(engineEngine,
&recorderObject, &audioSrc,
        &audioSnk, 1, id, req);
    result = (*recorderObject)->Realize(recorderObject, SL_
BOOLEAN_FALSE);
    result = (*recorderObject)->GetInterface(recorderObject, SL_
IID_RECORD, &recorderRecord);
    result = (*recorderObject)->GetInterface(recorderObject, SL_
IID_ANDROIDSIMPLEBUFFERQUEUE, &recorderBufferQueue);
    result = (*recorderBufferQueue)->RegisterCallback(recorderBuffe
rQueue, bqRecorderCallback, NULL);
    return JNI_TRUE;
}
```

`startRecording` enqueues the buffer to store the recording audio and set the audio object state as recording:

```
void startRecording() {
    SLresult result;
    recordF = fopen("/sdcard/test.pcm", "wb");
```

```
    result = (*recorderRecord)->SetRecordState(recorderRecord, SL_
RECORDSTATE_STOPPED);
    result = (*recorderBufferQueue)->Clear(recorderBufferQueue);
    recordCnt = 0;
    result = (*recorderBufferQueue)->Enqueue(recorderBufferQueue,
recorderBuffer,
            RECORDER_FRAMES * sizeof(short));
    result = (*recorderRecord)->SetRecordState(recorderRecord, SL_
RECORDSTATE_RECORDING);
}
```

Every time the buffer queue is ready to accept a new data block, the
bqRecorderCallback callback method is invoked. This happens when a buffer is
filled with audio data:

```
void bqRecorderCallback(SLAndroidSimpleBufferQueueItf bq, void
*context) {
    int numOfRecords = fwrite(recorderBuffer, sizeof(short),
RECORDER_FRAMES, recordF);
    fflush(recordF);
    recordCnt++;
    SLresult result;
    if (recordCnt*5 < RECORD_TIME) {
     result = (*recorderBufferQueue)->Enqueue(recorderBufferQueue,
recorderBuffer,
        RECORDER_FRAMES * sizeof(short));
    } else {
    result = (*recorderRecord)->SetRecordState(recorderRecord, SL_
RECORDSTATE_STOPPED);
    if (SL_RESULT_SUCCESS == result) {
      fclose(recordF);
    }
    }
}
```

play.c contains the code to create an audio player object and play the audio.

createBufferQueueAudioPlayer creates and realizes an audio player object
which plays audio from the buffer queue:

```
void createBufferQueueAudioPlayer() {
    SLresult result;
    SLDataLocator_AndroidSimpleBufferQueue loc_bufq = {SL_
DATALOCATOR_ANDROIDSIMPLEBUFFERQUEUE, 1};
    SLDataFormat_PCM format_pcm = {SL_DATAFORMAT_PCM, 1, SL_
SAMPLINGRATE_16,
        SL_PCMSAMPLEFORMAT_FIXED_16, SL_PCMSAMPLEFORMAT_FIXED_16,
        SL_SPEAKER_FRONT_CENTER, SL_BYTEORDER_LITTLEENDIAN};
```

```
    SLDataSource audioSrc = {&loc_bufq, &format_pcm};
    SLDataLocator_OutputMix loc_outmix = {SL_DATALOCATOR_OUTPUTMIX,
outputMixObject};
    SLDataSink audioSnk = {&loc_outmix, NULL};
    const SLInterfaceID ids[3] = {SL_IID_BUFFERQUEUE, SL_IID_
EFFECTSEND, SL_IID_VOLUME};
    const SLboolean req[3] = {SL_BOOLEAN_TRUE, SL_BOOLEAN_TRUE, SL_
BOOLEAN_TRUE};
    result = (*engineEngine)->CreateAudioPlayer(engineEngine,
&bqPlayerObject, &audioSrc, &audioSnk, 3, ids, req);
    result = (*bqPlayerObject)->Realize(bqPlayerObject, SL_BOOLEAN_
FALSE);
    result = (*bqPlayerObject)->GetInterface(bqPlayerObject, SL_
IID_PLAY, &bqPlayerPlay);
    result = (*bqPlayerObject)->GetInterface(bqPlayerObject, SL_
IID_BUFFERQUEUE,
            &bqPlayerBufferQueue);
    result = (*bqPlayerBufferQueue)->RegisterCallback(bqPlayerBuffe
rQueue, bqPlayerCallback, NULL);
    result = (*bqPlayerObject)->GetInterface(bqPlayerObject, SL_
IID_EFFECTSEND,
            &bqPlayerEffectSend);
    result = (*bqPlayerObject)->GetInterface(bqPlayerObject, SL_
IID_VOLUME, &bqPlayerVolume);
}
```

startPlaying fills the buffer with data from the test.cpm file and starts playing:

```
jboolean startPlaying() {
  SLresult result;
  recordF = fopen("/sdcard/test.pcm", "rb");
  noMoreData = 0;
  int numOfRecords = fread(recorderBuffer, sizeof(short),
RECORDER_FRAMES, recordF);
  if (RECORDER_FRAMES != numOfRecords) {
    if (numOfRecords <= 0) {
      return JNI_TRUE;
    }
    noMoreData = 1;
  }
result = (*bqPlayerBufferQueue)->Enqueue(bqPlayerBufferQueue,
recorderBuffer, RECORDER_FRAMES * sizeof(short));
  result = (*bqPlayerPlay)->SetPlayState(bqPlayerPlay, SL_
PLAYSTATE_PLAYING);
  return JNI_TRUE;
}
```

bqPlayerCallback every time the buffer queue is ready to accept a new buffer, this callback method is invoked. This happens when a buffer has finished playing:

```
void bqPlayerCallback(SLAndroidSimpleBufferQueueItf bq, void
*context) {
    if (!noMoreData) {
        SLresult result;
int numOfRecords = fread(recorderBuffer, sizeof(short), RECORDER_
FRAMES, recordF);
    if (RECORDER_FRAMES != numOfRecords) {
        if (numOfRecords <= 0) {
          noMoreData = 1;
          (*bqPlayerPlay)->SetPlayState(bqPlayerPlay, SL_PLAYSTATE_
STOPPED);
          fclose(recordF);
          return;
        }
        noMoreData = 1;
    }
    result = (*bqPlayerBufferQueue)->Enqueue(bqPlayerBufferQueue,
recorderBuffer,  RECORDER_FRAMES * sizeof(short));
    } else {
        (*bqPlayerPlay)->SetPlayState(bqPlayerPlay, SL_PLAYSTATE_
STOPPED);
        fclose(recordF);
    }
}
```

OpenSLESDemo.cpp contains the code to create the OpenSL ES engine object, free the objects, and register the native methods:

naCreateEngine creates the engine object and outputs the mix object.

```
void naCreateEngine(JNIEnv* env, jclass clazz) {
    SLresult result;
    result = slCreateEngine(&engineObject, 0, NULL, 0, NULL, NULL);
    result = (*engineObject)->Realize(engineObject, SL_BOOLEAN_
FALSE);
    result = (*engineObject)->GetInterface(engineObject, SL_IID_
ENGINE, &engineEngine);
    const SLInterfaceID ids[1] = {SL_IID_ENVIRONMENTALREVERB};
    const SLboolean req[1] = {SL_BOOLEAN_FALSE};
    result = (*engineEngine)->CreateOutputMix(engineEngine,
&outputMixObject, 1, ids, req);
    result = (*outputMixObject)->Realize(outputMixObject, SL_
BOOLEAN_FALSE);
    result = (*outputMixObject)->GetInterface(outputMixObject, SL_
IID_ENVIRONMENTALREVERB,
```

```
        &outputMixEnvironmentalReverb);
    if (SL_RESULT_SUCCESS == result) {
        result = (*outputMixEnvironmentalReverb)->SetEnvironmental
ReverbProperties(
            outputMixEnvironmentalReverb, &reverbSettings);
    }
}
```

5. Add the following permissions to the `AndroidManifest.xml` file.

```
<uses-permission android:name="android.permission.RECORD_AUDIO"/>
<uses-permission android:name="android.permission.WRITE_EXTERNAL_
STORAGE"/>
<uses-permission android:name="android.permission.MODIFY_AUDIO_
SETTINGS"></uses-permission>
```

6. Add an `Android.mk` file in the `jni` folder with the following content:

```
LOCAL_PATH := $(call my-dir)
include $(CLEAR_VARS)
LOCAL_MODULE    := OpenSLESDemo
LOCAL_SRC_FILES := OpenSLESDemo.cpp record.c play.c
LOCAL_LDLIBS := -llog
LOCAL_LDLIBS    += -lOpenSLES
include $(BUILD_SHARED_LIBRARY)
```

7. Build and run the Android project, and use the following command to monitor the `logcat` output:

```
$ adb logcat -v time OpenSLESDemo:I *:S
```

8. The application GUI is shown in the following screenshot:

    ❏   We can start the audio recording by clicking on the **Record** button. The recording will last for 15 seconds. The `logcat` output will be as shown in the following screenshot:

```
10-26 13:41:48.502 I/OpenSLESDemo(32672): recorder started recording
10-26 13:41:53.497 I/OpenSLESDemo(32672): write 80000
10-26 13:41:58.492 I/OpenSLESDemo(32672): write 80000
10-26 13:42:03.477 I/OpenSLESDemo(32672): write 80000
10-26 13:42:04.207 I/OpenSLESDemo(32672): recorder stopped
```

    ❏   Once the recording is finished. There will be a `/sdcard/test.pcm` file created at the Android device. We can click on the **Play** button to play the audio file. The `logcat` output will be as shown in the following screenshot:

```
10-26 13:45:35.804 I/OpenSLESDemo(32672): player created
10-26 13:45:35.804 I/OpenSLESDemo(32672): read 80000
10-26 13:45:35.804 I/OpenSLESDemo(32672): player started playing
10-26 13:45:40.769 I/OpenSLESDemo(32672): read 80000
10-26 13:45:45.774 I/OpenSLESDemo(32672): read 80000
10-26 13:45:50.778 I/OpenSLESDemo(32672): no data for playing
10-26 13:45:50.778 I/OpenSLESDemo(32672): player finished playing
```

## How it works...

This sample project demonstrates how to use OpenSL ES Audio library. We will first explain some key concepts and then describe how we used the recording and playback API.

### Object creation

An object does not have an actual representation in code and the creation of an object is done through interface. Every method which creates an object returns a `SLObjectInf` interface, which can be used to perform the basic operations on the object and access other interfaces of the object. The steps for object creation is described as follows:

1. Create an engine object. The engine object is the entry point of OpenSL ES API. Creating an engine object is done with the global function `slCreateEngine()`, which returns a `SLObjectItf` interface.

2. Realize the engine object. An object cannot be used until it is realized. We will discuss this in detail in the following section.

3. Obtain the `SLEngineItf` interface of the engine object through the `GetInterface()` method of the `SLObjectItf` interface.

4. Call the object creation method provided by the `SLEngineItf` interface. A `SLObjectItf` interface of the newly created object is returned upon success.

5. Realize the newly created object.

6. Manipulate the created objects or access other interfaces through the `SLObjectItf` interface of the object.

7. After you are done with the object, call the `Destroy()` method of the `SLObjectItf` interface to free the object and its resources.

In our sample project, we created and realized the engine object, and obtained the `SLEngineItf` interface at the `naCreateEngine` function of `OpenSLESDemo.cpp`. We then called the `CreateAudioRecorder()` method, exposed by the `SLEngineItf` interface, to create an audio recorder object at `createAudioRecorder` function of `record.c`. In the same function, we also realized the audio recorder object and accessed a few other interfaces of the object through the `SLObjectItf` interface returned at object creation. After we are finished with the recorder object, we called the `Destroy()` method to free the object and its resources, as shown in the `naShutdown` function of `OpenSLESDemo.cpp`.

One more thing to take note of on object creation is the interface request. An object creation method normally accepts three parameters related to interfaces, as shown in the `CreateAudioPlayer` method of the `SLEngineItf` interface as shown in the following code snippet:

```
SLresult (*CreateAudioPlayer) (
SLEngineItf self,
SLObjectItf * pPlayer,
SLDataSource *pAudioSrc,
SLDataSink *pAudioSnk,
SLuint32 numInterfaces,
const SLInterfaceID * pInterfaceIds,
const SLboolean * pInterfaceRequired
);
```

The last three input arguments are related to interfaces. The `numInterfaces` argument indicates the number of interfaces we request to access. `pInterfaceIds` is an array of the `numInterfaces` interface IDs, which indicates the interface types the object should support. `pInterfaceRequired` is an array of `SLboolean`, specifying whether the requested interface is optional or required. In our audio player example, we called the `CreateAudioPlayer` method to request three types of interfaces (`SLAndroidSimpleBufferQueueItf`, `SLEffectSendItf`, and `SLVolumeItf` indicated by `SL_IID_BUFFERQUEUE`, `SL_IID_EFFECTSEND`, and `SL_IID_VOLUME` respectively). Since all elements of the `req` array are `true`, all the interfaces are required. If the object cannot provide any of the interfaces, the object creation will fail:

```
const SLInterfaceID ids[3] = {SL_IID_BUFFERQUEUE, SL_IID_EFFECTSEND,
SL_IID_VOLUME};
const SLboolean req[3] = {SL_BOOLEAN_TRUE, SL_BOOLEAN_TRUE,  SL_
BOOLEAN_TRUE};
result = (*engineEngine)->CreateAudioPlayer(engineEngine,
&bqPlayerObject, &audioSrc, &audioSnk, 3, ids, req);
```

Note that an object can have implicit and explicit interfaces. The implicit interfaces are available for every object of the type. For example, the `SLObjectItf` interface is an implicit interface for all objects of all types. It is not necessary to request the implicit interfaces in the object creation method. However, if we want to access some explicit interfaces, we must request them in the method.

For more information on interfaces refer to *Section 3.1.6, The Relationship Between Objects and Interfaces* in the *OpenSL ES 1.0.1 Specification* document.

## Changing states of objects

The object creation method creates an object and puts it in an unrealized state. At this state, the resources of the object have not been allocated, therefore it is not usable.

We will need to call the `Realize()` method of the `SLObjectItf` interface of the object to cause the object to transit to the realized state, where the resources are allocated and the interfaces can be accessed.

Once we are done with the object, we call the `Destroy()` method to free the object and its resources. This call internally transfers the object through the unrealized stage, where the resources are freed. Therefore, the resources are freed first before the object itself.

In this recipe, we illustrate the recording and playback APIs with our sample project.

## Use and build with OpenSL ES Audio library

In order to call the API functions, we must add the following lines to our code:

```
#include <SLES/OpenSLES.h>
```

If we are using Android-specific features as well, we should include another header:

```
#include <SLES/OpenSLES_Android.h>
```

In the `Android.mk` file, we must add the following line to link to the native OpenSL ES Audio library:

```
LOCAL_LDLIBS += libOpenSLES
```

## OpenSL ES audio recording

Because the MIME data format and the `SLAudioEncoderItf` interface are not available for the audio recorder on Android, we can only record audio in the PCM format. Our example demonstrates how to record audio in the PCM format and save the data into a file. This can be illustrated using the following diagram:

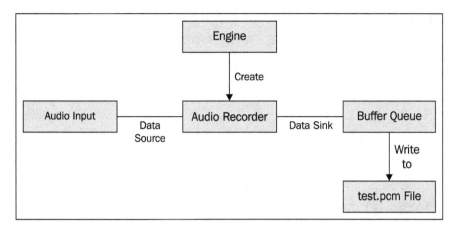

At the `createAudioRecorder` function of `record.c`, we create and realize an audio recorder object. We set the audio input as data source, and an Android buffer queue as data sink. Note that we registered the `bqRecorderCallback` function as the callback function for buffer queue. Whenever the buffer queue is ready for a new buffer, the `bqRecorderCallback` function will be called to save the buffer data to the `test.cpm` file and enqueue the buffer again for recording new audio data. At the `startRecording` function, we start the recording.

The callback functions in OpenSL ES are executed from internal non-application threads. The threads are not managed by Dalvik VM and therefore they cannot access JNI. These threads are critical to the integrity of the OpenSL ES implementation, so the callback functions should not block or perform any heavy-processing tasks.

In case we need to perform heavy tasks when the callback function is triggered, we should post an event for another thread to process such tasks.

This also applies to the OpenMAX AL library that we are going to cover in next recipe. More detailed information can be obtained from the NDK OpenSL ES documentation at the `docs/opensles/` folder.

## OpenSL ES audio playback

Android OpenSL ES library provides lots of features for audio playback. We can play encoded audio files, including mp3, aac, and so on. Our example shows how to play the PCM audio. This can be illustrated as shown in the following diagram:

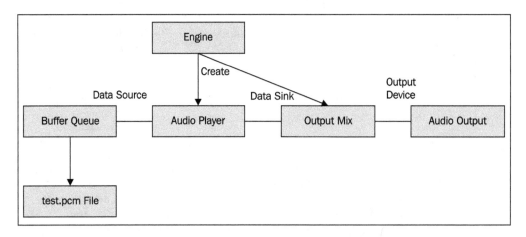

We created and realized the engine object and the output mix object in the naCreateEngine function in OpenSLESDemo.cpp. The audio player object is created in the createBufferQueueAudioPlayer function of play.c with an Android buffer queue as data source and the output mix object as data sink. The bqPlayerCallback function is registered as the callback method through a SLAndroidSimpleBufferQueueItf interface. Whenever the player finishes playing a buffer, the buffer queue is ready for new data and the callback function bqPlayerCallback will be invoked. The method reads data from the test.pcm file into the buffer and enqueues it.

In the startPlaying function, we read the initial data into the buffer and set the player state to SL_PLAYSTATE_PLAYING.

## There's more...

OpenSL ES is a complex library with a more than 500 page long specification. The specification is a good reference when developing applications with OpenSL ES and it is available with the Android NDK.

The Android NDK also comes with a native-audio example, which demonstrates usage of a lot more OpenSL ES functions.

# Programming with the OpenMAX AL multimedia library in Android NDK

OpenMAX AL is an application-level multimedia library in C. Android NDK multimedia APIs are based on the OpenMAX AL 1.0.1 standard with Android-specific extensions. The API is available for Android 4.0 or higher. We should note that the API is evolving and the Android NDK team mentioned that the future version of OpenMAX AL API may require developers to change their code.

## Getting ready...

Before we start coding with the OpenMAX AL library, it is important to understand some basics about the library. We will briefly describe the library in the following text.

OpenMAX AL refers to the Application Layer interface of the **Open Media Acceleration** (**OpenMAX**) library. It is a royalty-free, cross-platform, C-language application level API for developers to create multimedia applications. Its main features include media recording, media playback, media controls (for example, brightness control), and effects. Compared to OpenSL ES library, OpenMAX AL provides features for both video and audio, but it lacks certain audio features like 3D audio and audio effects which OpenSL ES can provide. Some applications may need to use both libraries.

OpenMAX AL defines two profiles, namely media playback and media player/recorder. Android does not implement all features required by either profile, therefore the OpenMAX AL library in Android does not conform either profile. In addition, Android implements some features specific to Android.

The main features provided by Android OpenMAX AL implementation is the ability to process the MPEG-2 transport stream. We can demultiplex the stream, decode the video and audio, and render them as audio output or to the phone screen. This library allows us to have complete control over the media data before it is passed for presentation. For example, we can call OpenGL ES functions to apply graphics effect on video data before rendering it.

For a detailed description of what is supported on Android, we can refer to the OpenMAX AL for Android documentation available with the Android NDK under the docs/openmaxal/ folder.

The design of OpenMAX AL library is similar to OpenSL ES library. They both adopt an object-oriented approach and the fundamental concepts including objects and interfaces are the same. Readers should refer to the previous recipe for a detailed explanation on these concepts.

## How to do it...

The following steps describe how to create a simple Android video playback application using the OpenMAX AL functions:

1. Create an Android application named `OpenMAXSLDemo`. Set the package name as `cookbook.chapter7.openmaxsldemo`. Refer to the *Loading native libraries and registering native methods* recipe of *Chapter 2, Java Native Interface* for more detailed instructions.

2. Right-click on the project **OpenMAXSLDemo**, select **Android Tools | Add Native Support**.

3. Add a Java file named `MainActivity.java` in the package `cookbook.chapter7.openmaxsldemo`. This Java file loads the native library `OpenMAXSLDemo`, sets the view, and calls the native methods to play the video.

4. Add the `mylog.h` and `OpenMAXSLDemo.c` files in the `jni` folder. A part of the code in `OpenMAXSLDemo.c` is showed in the following code snippet.

    `naCreateEngine` creates and realizes the engine object and the output mix object.

    ```
    void naCreateEngine(JNIEnv* env, jclass clazz) {
        XAresult res;
        res = xaCreateEngine(&engineObject, 0, NULL, 0, NULL, NULL);
        res = (*engineObject)->Realize(engineObject, XA_BOOLEAN_FALSE);
        res = (*engineObject)->GetInterface(engineObject, XA_IID_
    ENGINE, &engineEngine);
        res = (*engineEngine)->CreateOutputMix(engineEngine,
    &outputMixObject, 0, NULL, NULL);
        res = (*outputMixObject)->Realize(outputMixObject, XA_BOOLEAN_
    FALSE);
    }
    ```

    `naCreateStreamingMediaPlayer` creates and realizes a media player object with the data source and data sink. It obtains the buffer queue interface and registers the `AndroidBufferQueueCallback` function as the callback function. The callback function will be invoked after a buffer is processed:

    ```
    jboolean naCreateStreamingMediaPlayer(JNIEnv* env, jclass clazz,
    jstring filename) {
        XAresult res;
        const char *utf8FileName = (*env)->GetStringUTFChars(env,
    filename, NULL);
        file = fopen(utf8FileName, "rb");
        XADataLocator_AndroidBufferQueue loc_abq = { XA_DATALOCATOR_
    ANDROIDBUFFERQUEUE, NB_BUFFERS };
        XADataFormat_MIME format_mime = {XA_DATAFORMAT_MIME, XA_
    ANDROID_MIME_MP2TS, XA_CONTAINERTYPE_MPEG_TS };
        XADataSource dataSrc = {&loc_abq, &format_mime};
    ```

```
    XADataLocator_OutputMix loc_outmix = { XA_DATALOCATOR_
OUTPUTMIX, outputMixObject };
    XADataSink audioSnk = { &loc_outmix, NULL };
    XADataLocator_NativeDisplay loc_nd = {XA_DATALOCATOR_
NATIVEDISPLAY,
             (void*)theNativeWindow, NULL};
    XADataSink imageVideoSink = {&loc_nd, NULL};
    XAboolean required[NB_MAXAL_INTERFACES] = {XA_BOOLEAN_TRUE, XA_
BOOLEAN_TRUE};
    XAInterfaceID iidArray[NB_MAXAL_INTERFACES] = {XA_IID_PLAY, XA_
IID_ANDROIDBUFFERQUEUESOURCE};
    res = (*engineEngine)->CreateMediaPlayer(engineEngine,
&playerObj, &dataSrc, NULL,   &audioSnk, &imageVideoSink, NULL,
NULL, NB_MAXAL_INTERFACES, iidArray, required );
    (*env)->ReleaseStringUTFChars(env, filename, utf8FileName);
    res = (*playerObj)->Realize(playerObj, XA_BOOLEAN_FALSE);
    res = (*playerObj)->GetInterface(playerObj, XA_IID_PLAY,
&playerPlayItf);
    res = (*playerObj)->GetInterface(playerObj, XA_IID_
ANDROIDBUFFERQUEUESOURCE, &playerBQItf);
    res = (*playerBQItf)->SetCallbackEventsMask(playerBQItf, XA_
ANDROIDBUFFERQUEUEEVENT_PROCESSED);
    res = (*playerBQItf)->RegisterCallback(playerBQItf,
AndroidBufferQueueCallback, NULL);
    if (!enqueueInitialBuffers(JNI_FALSE)) {
        return JNI_FALSE;
    }
    res = (*playerPlayItf)->SetPlayState(playerPlayItf, XA_
PLAYSTATE_PAUSED);
    res = (*playerPlayItf)->SetPlayState(playerPlayItf, XA_
PLAYSTATE_PLAYING);
    return JNI_TRUE;
}
```

AndroidBufferQueueCallback is the callback function registered to refill the buffer with media data or handle commands:

```
XAresult AndroidBufferQueueCallback(XAAndroidBufferQueueI
tf caller, void *pCallbackContext, void *pBufferContext,  void
*pBufferData, XAuint32 dataSize,  XAuint32 dataUsed, const
XAAndroidBufferItem *pItems, XAuint32 itemsLength) {
    XAresult res;
    int ok;
    ok = pthread_mutex_lock(&mutex);
    if (discontinuity) {
        if (!reachedEof) {
            res = (*playerBQItf)->Clear(playerBQItf);
```

```
            rewind(file);
              (void) enqueueInitialBuffers(JNI_TRUE);
        }
        discontinuity = JNI_FALSE;
        ok = pthread_cond_signal(&cond);
        goto exit;
    }
    if ((pBufferData == NULL) && (pBufferContext != NULL)) {
        const int processedCommand = *(int *)pBufferContext;
        if (kEosBufferCntxt == processedCommand) {
            goto exit;
        }
    }
    if (reachedEof) {
        goto exit;
    }
    size_t nbRead;
    size_t bytesRead;
    bytesRead = fread(pBufferData, 1, BUFFER_SIZE, file);
    if (bytesRead > 0) {
        if ((bytesRead % MPEG2_TS_PACKET_SIZE) != 0) {
            LOGI(2, "Dropping last packet because it is not
whole");
        }
        size_t packetsRead = bytesRead / MPEG2_TS_PACKET_SIZE;
        size_t bufferSize = packetsRead * MPEG2_TS_PACKET_SIZE;
        res = (*caller)->Enqueue(caller, NULL, pBufferData,
bufferSize, NULL, 0);
    } else {
        XAAndroidBufferItem msgEos[1];
        msgEos[0].itemKey = XA_ANDROID_ITEMKEY_EOS;
        msgEos[0].itemSize = 0;
        res = (*caller)->Enqueue(caller, (void *)&kEosBufferCntxt,
NULL, 0, msgEos, sizeof(XAuint32)*2);
        reachedEof = JNI_TRUE;
    }
exit:
    ok = pthread_mutex_unlock(&mutex);
    return XA_RESULT_SUCCESS;
}
```

5. Add an `Android.mk` file in the `jni` folder with the following content:

```
LOCAL_PATH := $(call my-dir)
include $(CLEAR_VARS)
LOCAL_MODULE     := OpenMAXSLDemo
LOCAL_SRC_FILES := OpenMAXSLDemo.c
LOCAL_LDLIBS := -llog
LOCAL_LDLIBS     += -landroid
LOCAL_LDLIBS     += -lOpenMAXAL
include $(BUILD_SHARED_LIBRARY)
```

6. We can use the `NativeMedia.ts` video file available in the `samples/native-media/` directory for testing. The following command can be used to put the video file into the `/sdcard/` directory of the testing Android device:

```
$ adb push NativeMedia.ts /sdcard/
```

7. Build and start the Android application. We can see the GUI as shown in the following screenshot:

We can press **Play** to start playing the video.

## How it works...

In this recipe, we used the OpenMAX AL library to implement a simple video player.

### Use and build with the OpenMAX AL multimedia library:

In order to call the API functions, we must add the following line to our code:

```
#include <OMXAL/OpenMAXAL.h>
```

If we are also using Android-specific features, we should include another header:

```
#include <OMXAL/OpenMAXAL_Android.h>
```

In the `Android.mk` file, we must add the following line to link to the OpenMAX AL multimedia library:

```
LOCAL_LDLIBS += libOpenMAXAL
```

## OpenMAX AL video playback

Our sample project that is a simplified version of the native media project comes with the Android NDK. The following diagram illustrates how the application works:

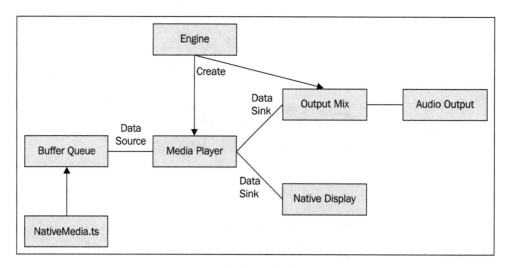

In our code, we created and realized the engine and output mix objects at `naCreateEngine` function. At the `naCreateStreamingMediaPlayerfunction` function, we created and realized the media player object with the audio data sink set as output mix, video data sink set as native display, and data source set as the Android buffer queue.

When a buffer is consumed, the callback function `AndroidBufferQueueCallback` is invoked, where we refill the buffer with data from the `NativeMedia.ts` file and enqueue it to the buffer queue.

## There's more....

OpenMAX AL is a complex library. The specification is a good reference when developing applications with OpenMAX AL and it is available with the Android NDK. The Android NDK also comes with a native-media example, which is a good example of how to use the API.

# 8
# Porting and Using the Existing Libraries with Android NDK

In this chapter, we will cover the following recipes:

- ▶ Porting a library as a shared library module with the Android NDK build system
- ▶ Porting a library as a static library module with the Android NDK build system
- ▶ Porting a library with its existing build system using the Android NDK toolchain
- ▶ Using a library as a prebuilt library
- ▶ Using a library in multiple projects with import-module
- ▶ Porting a library that requires RTTI, exception, and STL support

## Introduction

There are many C/C++ libraries for the Desktop computing world. These libraries can save us a huge amount of effort if we can reuse them on an Android platform. Android NDK makes this possible. In this chapter, we will discuss how to port the existing libraries to Android with NDK.

We will first introduce how to build libraries with the Android NDK build system. We can build a library as a static library module or a shared library module. The differences between the two will be discussed in this chapter.

We can also use the Android NDK toolchain as a standalone cross compiler, which is covered next. We will then describe how to use the compiled libraries as prebuilt modules.

We often use the same library in multiple Android projects. We can use the **Import-module** feature to link to the same library module to multiple projects while maintaining a single copy of the library.

Many C++ libraries require STL, C++ exceptions, and **Run-Time Type Information** (**RTTI**) supports, which are not available with the Android default C++ runtime library. We will illustrate how to enable these supports by using the popular boost library as an example.

# Porting a library as a shared library module with the Android NDK build system

This recipe will discuss how to port an existing library as a shared library with the Android NDK build system. We will use the open source libbmp library as an example.

## Getting ready

Readers are recommended to read the *Building an Android NDK application at the command line* recipe in *Chapter 3, Build and Debug NDK Applications*, before going through this one.

## How to do it...

The following steps describe how to create our sample Android project that demonstrates porting the libbmp library as a shared library:

1. Create an Android application named PortingShared with native support. Set the package name as cookbook.chapter8.portingshared. Please refer to the *Loading native libraries and registering native methods* recipe of *Chapter 2, Java Native Interface*, if you want more detailed instructions.

2. Add a Java file MainActivity.java under the cookbook.chapter8.portingshared package. This Java file simply loads the shared library .bmp and PortingShared, and calls the native method naCreateABmp.

3. Download the libbmp library from http://code.google.com/p/libbmp/downloads/list, and extract the archive file. Create a folder named libbmp under the jni folder, and copy the src/bmpfile.c and src/bmpfile.h files from the extracted folder to the libbmp folder.

4. Remove the following code from bmpfile.h if you are using NDK r8 and below:

```
#ifndef uint8_t
typedef unsigned char uint8_t;
#endif
#ifndef uint16_t
typedef unsigned short uint16_t;
```

```
#endif
#ifndef uint32_t
typedef unsigned int uint32_t;
#endif
```

5.  Then, add the following line of code:

```
#include <stdint.h>
```

> The code changes for `bmpfile.h` are only necessary for Android NDK r8 and below. Compiling the library will return an error `"error: redefinition of typedef 'uint8_t'"`. This is a bug in the NDK build system as the `uint8_t` definition is enclosed by the `#ifndef` preprocessor. It has been fixed since NDK r8b, and we don't need to change the code if we're using r8b and above.

6.  Create an `Android.mk` file under the `libbmp` folder to compile `libbmp` as a shared library `libbmp.so`. The content of this `Android.mk` file is as follows:

```
LOCAL_PATH := $(call my-dir)
include $(CLEAR_VARS)
LOCAL_MODULE     := libbmp
LOCAL_SRC_FILES := bmpfile.c
include $(BUILD_SHARED_LIBRARY)
```

7.  Create another folder named `libbmptest` under the `jni` folder. Add the `mylog.h` and `PortingShared.c` files under it. `PortingShared.c` implements the native method `naCreateABmp`, which uses functions defined in the `libbmp` library to create a bitmap image and save it to `/sdcard/test_shared.bmp`. You will need to change directory if the `/sdcard` directory is not available on your devices:

```
void naCreateABmp(JNIEnv* env, jclass clazz, jint width, jint
height, jint depth) {
  bmpfile_t *bmp;
  int i, j;
  rgb_pixel_t pixel = {128, 64, 0, 0};
  for (i = 10, j = 10; j < height; ++i, ++j) {
    bmp_set_pixel(bmp, i, j, pixel);
    pixel.red++;
    pixel.green++;
    pixel.blue++;
    bmp_set_pixel(bmp, i + 1, j, pixel);
    bmp_set_pixel(bmp, i, j + 1, pixel);
  }
  bmp_save(bmp, "/sdcard/test_shared.bmp");
  bmp_destroy(bmp);
}
```

8. Create another `Android.mk` file under the `libbmptest` folder to compile the `PortingShared.c` file as another shared library `libPortingShared.so`. The content of this `Android.mk` file is as follows:

```
LOCAL_PATH := $(call my-dir)
include $(CLEAR_VARS)
LOCAL_MODULE    := PortingShared
LOCAL_C_INCLUDES := $(LOCAL_PATH)/../libbmp/
LOCAL_SRC_FILES := PortingShared.c
LOCAL_SHARED_LIBRARIES := libbmp
LOCAL_LDLIBS := -llog
include $(BUILD_SHARED_LIBRARY)
```

9. Create an `Android.mk` file under the `jni` folder with the following content:

```
LOCAL_PATH := $(call my-dir)
include $(call all-subdir-makefiles)
```

10. Add the `WRITE_EXTERNAL_STORAGE` permission to the `AndroidManifest.xml` file as follows:

```
<uses-permission android:name="android.permission.WRITE_EXTERNAL_STORAGE"/>
```

11. Build and run the Android project. A bitmap file `test_shared.bmp` should be created at the `sdcard` folder of the Android device. We can use the following command to get the file:

```
$ adb pull /sdcard/test_shared.bmp .
```

The following is a `.bmp` file:

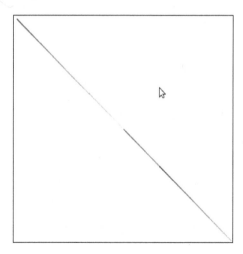

## How it works...

The sample project demonstrates how to port the `libbmp` code as a shared library and use it in the native code `PortingShared.c`.

**Shared library**: A shared library can be shared by multiple executables and libraries. The Android native code is usually compiled as shared libraries and loaded by the Java code. In fact, the Android build system only packages shared libraries into the application's `apk` file. Therefore, we must provide at least one shared library to contain our native code.

 We can still use static libraries to generate shared libraries, as we will see in the *Porting a library as static library module with Android NDK build system* recipe.

Our sample project builds two shared libraries, namely `libbmp.so` and `libPortingShared.so`. We can find these libraries under the `libs` folder of the project. `libPortingShared.so` depends on `libbmp.so`, since `PortingShared.c` calls functions defined in the `libbmp` library.

In our Java file, we need to load `libbmp.so` before `libPortingShared.so`, as follows:

```
static {
        System.loadLibrary("bmp");
          System.loadLibrary("PortingShared");
}
```

**Understand the Android.mk files**: Android NDK provides an easy-to-use build system, which frees us from writing makefiles. However, we still need to provide some basic inputs to the system through `Android.mk` and `Application.mk`. We only discuss `Android.mk` in this recipe.

The `Android.mk` file is a GNU makefile fragment that describes the sources to the Android build system. The sources are grouped into modules. Each module is a static or shared library. The Android NDK provides a few predefined variables and macros. Here, we will briefly describe the ones used in this recipe. We will introduce more predefined variables and macros in subsequent recipes and you can also refer to Android NDK `docs/ANDROID-MK.html` for more information.

- ▶ `CLEAR_VARS`: This variable points to a script, which undefines nearly all module description variables except `LOCAL_PATH`. We must include it before every new module, as follows:

  ```
  include $(CLEAR_VARS)
  ```

- ▶ `BUILD_SHARED_LIBRARY`: This variable points to a build script, which determines how to build a shared library from the sources listed, based on the module description. We must have `LOCAL_MODULE` and `LOCAL_SRC_FILES` defined when including this variable, as follows:

  ```
  include $(BUILD_SHARED_LIBRARY)
  ```

  Including it will generate a shared library `lib$(LOCAL_MODULE).so`.

- ▶ `my-dir`: This must be evaluated by using `$(call <macro>)`. The `my-dir` macro returns the path of the last included makefile, which is usually the directory containing the current `Android.mk` file. It is typically used to define the `LOCAL_PATH`, as follows:

  ```
  LOCAL_PATH := $(call my-dir)
  ```

- ▶ `all-subdir-makefiles`: This macro returns a list of `Android.mk` files located in all subdirectories of the current `my-dir` path. In our example, we used this macro in the `Android.mk` file under the `jni`, as follows:

  ```
  include $(call all-subdir-makefiles)
  ```

  This will include the two `Android.mk` files under `libbmp` and `libbmptest`.

- ▶ `LOCAL_PATH`: This is a module description variable, which is used to locate the path to the sources. It is usually used with the `my-dir` macro, as follows:

  ```
  LOCAL_PATH := $(call my-dir)
  ```

- ▶ `LOCAL_MODULE`: This is a module description variable, which defines the name of our module. Note that it must be unique among all module names and must not contain any space.

- ▶ `LOCAL_SRC_FILES`: This is a module description variable, which lists out the sources used to build the module. Note that the sources should be relative to `LOCAL_PATH`.

- ▶ `LOCAL_C_INCLUDES`: This is an optional module description variable, which provides a list of the paths that will be appended to the include search path at compilation. The paths should be relative to the NDK root directory. In `Android.mk`, under the `libbmptest` folder of our sample project, we used this variable as follows:

  ```
  LOCAL_C_INCLUDES := $(LOCAL_PATH)/../libbmp/
  ```

- ▶ `LOCAL_SHARED_LIBRARIES`: This is an optional module description variable, which provides a list of the shared libraries the current module depends on. In `Android.mk`, under the `libbmptest` folder of our sample project, we used this variable to include the `libbmp.so` shared library:

  ```
  LOCAL_SHARED_LIBRARIES := libbmp
  ```

▶ `LOCAL_LDLIBS`: This is an optional module description variable, which provides a list of linker flags. It is useful to pass the system libraries with the `-l` prefix. In our sample project, we used it to link the system log library:

```
LOCAL_LDLIBS := -llog
```

With the preceding description, it is now fairly easy to understand the three `Android.mk` files used in our sample project. `Android.mk` under `jni` simply includes another two `Android.mk` files. `Android.mk` under the `libbmp` folder compiles the `libbmp` sources as a shared library `libbmp.so`, and `Android.mk` under the `libbmptest` folder compiles `PortingShared.c` as the `libPortingShared.so` shared library, which depends upon the `libbmp.so` library.

## See also

It is possible to use a shared library in the native code, as we have demonstrated in the *Programming with dynamic linker library at Android NDK* recipe in *Chapter 6, Other Android NDK API*.

# Porting a library as a static library module with the Android NDK build system

The previous recipe discussed how to port a library as a shared library module with the `libbmp` library as an example. In this recipe, we will demonstrate how to port the `libbmp` library as a static library.

## Getting ready

Readers are recommended to read the *Building an Android NDK application at the command line* recipe in *Chapter 3, Build and Debug NDK Applications*, before going through this one.

## How to do it...

The following steps describe how to create our sample Android project that demonstrates porting the `libbmp` library as a static library:

1. Create an Android application named `PortingStatic` with native support. Set the package name as `cookbook.chapter8.portingstatic`. Please refer to the *Loading native libraries and registering native methods* recipe of *Chapter 2, Java Native Interface*, if you want more detailed instructions.

2. Add a Java file `MainActivity.java` under the `cookbook.chapter8.portingstatic` package. This Java file simply loads the shared library `PortingStatic`, and calls the native method `naCreateABmp`.

3. Follow step 3 of the *Porting a library as shared library module with the Android NDK build system* recipe to download the `libbmp` library and make changes.

4. Create an `Android.mk` file under the `libbmp` folder to compile `libbmp` as a static library `libbmp.a`. The content of this `Android.mk` file is as follows:

```
LOCAL_PATH := $(call my-dir)
include $(CLEAR_VARS)
LOCAL_MODULE    := libbmp
LOCAL_SRC_FILES := bmpfile.c
include $(BUILD_STATIC_LIBRARY)
```

5. Create another folder `libbmptest` under the `jni` folder. Add the `mylog.h` and `PortingStatic.c` files to it. Note that the code for it is the same as the `naCreateABmp` method in previous chapter except that the `.bmp` file name is changed from `test_shared.bmp` to `test_static.bmp`.

6. Create another `Android.mk` file under the `libbmptest` folder to compile the `PortingStatic.c` file as a shared library `libPortingStatic.so`. The content of this `Android.mk` file is as follows:

```
LOCAL_PATH := $(call my-dir
include $(CLEAR_VARS)
LOCAL_MODULE    := PortingStatic
LOCAL_C_INCLUDES := $(LOCAL_PATH)/../libbmp/
LOCAL_SRC_FILES := PortingStatic.c
LOCAL_STATIC_LIBRARIES := libbmp
LOCAL_LDLIBS := -llog
include $(BUILD_SHARED_LIBRARY)
```

7. Create an `Android.mk` file under the `jni` folder with the following content:

```
LOCAL_PATH := $(call my-dir)
include $(call all-subdir-makefiles)
```

8. Add the `WRITE_EXTERNAL_STORAGE` permission to the `AndroidManifest.xml` file as follows:

```
<uses-permission android:name="android.permission.WRITE_EXTERNAL_
STORAGE"/>
```

9. Build and run the Android project. A bitmap file `test_static.bmp` should be created at the `sdcard` folder of the Android device. We can use the following command to get the file:

```
$ adb pull /sdcard/test_static.bmp .
```

This file is the same as the `test_static.bmp` file used in the previous recipe.

## How it works...

In the sample project, we build `libbmp` as a static library `libbmp.a`, which can be found under `obj/local/armeabi/` folder. We called the functions defined in `libbmp` in the native code `PortingStatic.c`.

A **static library** is simply an archive of object files compiled from the source code. They are built as files ending with ".a" suffix at Android NDK. A static library is copied into a targeted executable or library at build time by a compiler or linker. At Android NDK, static libraries are only used to build the shared libraries, because only shared libraries are packaged into the apk file for deployment.

Our sample project builds a static library `libbmp.a` and a shared library `libPortingStatic.so`. The `libPortingStatic.so` shared library is located under the `libs/armeabi` folder, and will be copied to the application's apk file. The `libbmp.a` library is used to build the `libPortingStatic.so` shared library. If you examine the symbols of the `libPortingStatic.so` library with the Eclipse project explorer, you will find that the symbols for the functions defined at `libbmp` are included. This is shown in the following screensnhot:

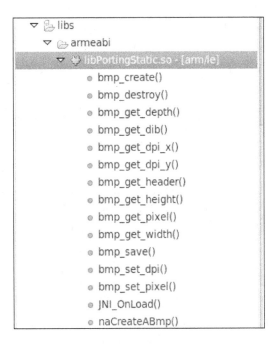

The functions `bmp_create`, `bmp_destroy`, and so on, are defined in `libbmp` and are included in the shared library `libPortingStatic.so`.

In our Java code, we will need to load the shared library with the following code:

```
static {
        System.loadLibrary("PortingStatic");
}
```

**Understand the Android.mk files**: The previous recipe already describes most of the predefined variables and macros used in the three `Android.mk` files in this recipe. Therefore, we only cover the ones that we have not seen in the previous recipe:

- ▸ `BUILD_STATIC_LIBRARY`: The variable points to a build script which will collect the information about the module and determine how to build a static library from the sources. The module built is usually listed in `LOCAL_STATIC_LIBRARIES` of another module. This variable is normally included in `Android.mk` as follows:

  ```
  include $(BUILD_STATIC_LIBRARY)
  ```

  In our sample project, we include this variable in the `Android.mk` file, under `jni/libbmp` folder.

- ▸ `LOCAL_STATIC_LIBRARIES`: This is a module description variable, which provides a list of static libraries the current module should be linked to. It only makes sense in shared library modules.

  In our project, we used this variable to link to the `libbmp.a` static library, as shown in the `Android.mk` file under the `jni/libbmptest/` folder.

  ```
  LOCAL_STATIC_LIBRARIES := libbmp
  ```

- ▸ `LOCAL_WHOLE_STATIC_LIBRARIES`: This is a variant of the `LOCAL_STATIC_LIBRARIES` variable. It indicates that the static libraries listed should be linked as whole archives. This will force all object files from the static libraries to be added to the current shared library module.

**Static versus shared**: Now that you have seen how to port an existing library as either a static or shared library, you may ask which one is better. The answer, as you might have expected, depends on our needs.

When you port a big library and only use a small portion of the functions provided by the library, then a static library is a good option. The Android NDK build system can resolve the dependencies at build time and only copy the parts that are used in the final shared library. This means a smaller library size and subsequently smaller `apk` file size.

 Sometimes, we need to force the entire static library to be built into the final shared library (for example, there are circular dependencies among several static libraries). We can use the `LOCAL_WHOLE_STATIC_LIBRARIES` variable at `Android.mk` or the `"--whole-archive"` linker flag.

When you port a library, which will be used by several Android apps, then shared library is a better choice. Suppose you want to build two Android apps, a video player and a video editor. Both apps will need a third-party `codec` library, which you can port to Android with NDK. In this case, you can port the library as a shared library in a separate `apk` file (for example, MX Player puts the `codecs` library in separate `apk` files) and the two apps can load the same library at runtime. This means that the users only need to download the library once to use both the apps.

Another case in which you may need the shared library is that a library `L` is used in multiple shared libraries. If `L` is a static library, each shared library will include a copy of its code and cause problems because of code duplication (for example, duplicated global variables).

## See also

We have actually ported a library as a static library with Android NDK build system before. Recall how we ported `libpng` as a static library in the _Managing assets at Android NDK_ recipe in _Chapter 5, Android Native Application API_.

# Porting a library with its existing build system using the Android NDK toolchain

The previous two recipes discussed how to port a library with the Android NDK build system. However, a lot of open source projects have their own build systems and sometimes it is troublesome to list out all sources in the `Android.mk` file. Fortunately, the Android NDK toolchain can also be used as a standalone cross compiler and we can use the cross compiler in an open source project's existing build system. This recipe will discuss how to port a library with its existing build system.

## How to do it...

The following steps describe how to create our sample project, which demonstrates porting the open source `libbmp` library with its existing build system:

1. Create an Android application named PortingWithBuildSystem with native support. Set the package name as `cookbook.chapter8.portingwithbuildsystem`. Please refer to the _Loading native libraries and registering native methods_ recipe of _Chapter 2, Java Native Interface_, if you want more detailed instructions.

2. Add a Java file `MainActivity.java` under the `cookbook.chapter8.portingwithbuildsystem` package. This Java file simply loads the shared library `PortingWithBuildSystem`, and calls the native method `naCreateABmp`.

3. Download the `libbmp` library from `http://code.google.com/p/libbmp/` `downloads/list` and extract the archive file to the `jni` folder. This will create a folder `libbmp-0.1.3` under the `jni` folder with the following content:

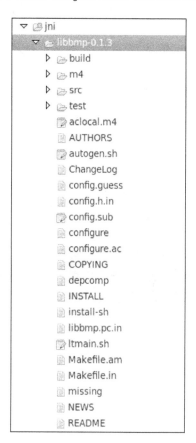

4. Follow step 3 of the *Porting a library as shared library module with Android NDK build system* recipe to update `src/bmpfile.h`.

5. Add a bash shell script file `build_android.sh` under the `libbmp-0.1.3` folder with the following content:

```
#!/bin/bash
NDK=<path to Android ndk folder>/android-ndk-r8b
SYSROOT=$NDK/platforms/android-8/arch-arm/
CFLAGS="-mthumb"
LDFLAGS="-Wl,--fix-cortex-a8"
export CC="$NDK/toolchains/arm-linux-androideabi-4.4.3/prebuilt/
linux-x86/bin/arm-linux-androideabi-gcc --sysroot=$SYSROOT"
./configure \
```

```
        --host=arm-linux-androideabi \
        --disable-shared \
        --prefix=$(pwd) \
        --exec-prefix=$(pwd)
    make clean
    make
    make install
```

6. Add the execute permission to the `build_android.sh` file with the following command:

    ```
    $ sudo chmod +x build_android.sh
    ```

7. At a command line shell, go to the `libbmp-0.1.3` directory, and enter the following command to build the library:

    ```
    $ ./build_android.sh
    ```

    The build will fail with the following errors:

    ```
    checking host system type... Invalid configuration `arm-linux-androideabi': system `an
    droideabi' not recognized
    configure: error: /bin/bash ./config.sub arm-linux-androideabi failed
    ```

    This is because the `config.guess` and `config.sub` scripts under the `libbmp-0.1.3` folder are out of date (the first line of these two files indicate that the timestamp is `2009-08-19`). We will need copies of the scripts with timestamps `2010-05-20` or later. The `config.guess` script can be found at `http://gcc.gnu.org/svn/gcc/branches/cilkplus/config.guess` and `config.sub` can be found at `http://gcc.gnu.org/svn/gcc/branches/cilkplus/config.sub`.

8. Try executing the `build_android.sh` script again. This time it finishes successfully. We should be able to find the `libbmp.a` static library under the `jni/libbmp-0.1.3/lib` folder and `bmpfile.h` under the `jni/libbmp-0.1.3/include` folder.

## How it works...

Many of the existing open source libraries can be built with the shell command "`./configure; make; make install`". In our sample project, we wrote a `build_android.sh` script to execute the three steps with the Android NDK cross compiler.

The following are a list of things we should consider when porting a library with the Android NDK cross compiler:

1. **Select the appropriate toolchain**: Based on the CPU architecture (ARM, x86 or MIPS) of our targeted devices, you need to choose the corresponding toolchain. The following toolchains are available under the `toolchains` folder of Android NDK r8d:

   □ **For ARM-based devices**: `arm-linux-androideabi-4.4.3`, `arm-linux-androideabi-4.6`, `arm-linux-androideabi-4.7`, and `arm-linux-androideabi-clang3.1`

   □ **For MIPS-based devices**: `mipsel-linux-android-4.4.3`, `mipsel-linux-android-4.6`, `mipsel-linux-android-4.7`, and `mipsel-linux-android-clang3.1`

   □ **For x86-based devices**: `x86-4.4.3`, `x86-4.6`, `x86-4.7`, and `x86-clang3.1`

   ▸ **Select the sysroot**: Based on the Android native API level and CPU architecture we want to target, you will need to choose the appropriate sysroot. The compiler will look for headers and libraries under the `sysroot` directory at compilation.

   The path to `sysroot` follows this format:

   `$NDK/platforms/android-<level>/arch-<arch>/`

   `$NDK` refers to the Android NDK root folder, `<level>` refers to the Android API level, and `<arch>` indicates the CPU architecture. In your `build_android.sh` script, SYSROOT is defined as follows:

   `SYSROOT=$NDK/platforms/android-8/arch-arm/`

2. **Specify the cross compiler**: The library's existing build system usually has a way for us to specify the cross compiler. It is usually through a configuration option or an environment variable.

   In `libbmp`, we can enter the "`./configure --help`" command to see how to set the compiler. The `compiler` command is specified through the environment variable CC, while the environment variables CFLAGS and LDFLAGS are used to specify the compiler flags and linker flags. In your `build_android.sh` script, these three environment variables are set as follows:

   ```
   export CFLAGS="-mthumb"
   export LDFLAGS="-Wl,--fix-cortex-a8"
   export CC="$NDK/toolchains/arm-linux-androideabi-4.4.3/prebuilt/
   linux-x86/bin/arm-linux-androideabi-gcc --sysroot=$SYSROOT"
   ```

The "-mthumb" compiler flag indicates that you will use the thumb instruction set rather than the ARM instruction set. The "-wl, --fix-cortex-a8" linker flag is required to route around a CPU bug in some Cortex-A8 implementations.

▶ **Specify the output locations for the header files and library binary**: You will usually want to place the library under jni/<library folder>/.

In the case of libbmp, the library binary is installed under the PREFIX/lib folder and the header file is installed under the EPREFIX/include folder. Therefore, we set PREFIX and EPREFIX to jni/libbmp-0.1.3 by passing the following options to configure the script:

```
--prefix=$(pwd) \
--exec-prefix=$(pwd)
```

▶ **Make and install the library**: You can simply execute "make; make install;" to build and install the library.

## There's more...

In your build_android.sh script, we have disabled shared library. If you remove the line "--disable-shared \", the build will generate both the shared library (libbmp.so) and the static library (libbmp.a) under the jni/libbmp-0.1.3/lib/ folder.

In your sample project, we used the NDK toolchain directly. This method has a serious limitation that you won't be able to use any C++ STL function, and C++ exceptions and RTTI are not supported. Android NDK actually allows you to create a customized toolchain installation with the script $NDK/build/tools/make-standalone-toolchain.sh. Suppose you're targeting Android API level 8; you can use the following command to install the toolchain at the /tmp/my-android-toolchain folder.

```
$ANDROID_NDK/build/tools/make-standalone-toolchain.sh
--platform=android-8 --install-dir=/tmp/my-android-toolchain
```

You can then use this toolchain by using the following commands:

```
export PATH=/tmp/my-android-toolchain/bin:$PATH
export CC=arm-linux-androideabi-gcc
```

Note that the installed toolchain will have a few libraries (`libgnustl_shared.so`, `libstdc++.a`, and `libsupc++.a`) under the `/tmp/my-android-toolchain/arm-linux-androideabi/lib/` folder. You can link against these libraries to enable exceptions, RTTI, and STL functions support. We will further discuss exception and STL support in the *Porting a library which requires RTTI* recipe.

More information about using the Android toolchain as a standalone compiler can be found at Android NDK in `docs/STANDALONE-TOOLCHAIN.html`.

# Using a library as a prebuilt library

The previous recipe described how to build an existing library with its own build system. We obtained a compiled static library `libbmp.a` of the open source `libbmp` library. This recipe will discuss how to use a prebuilt library.

## How to do it...

The following steps build an Android NDK application which uses prebuilt library. Note that the sample project is based on what we have done in the previous recipe. If you have not gone through previous recipe, you should do it now.

1. Open the `PortingWithBuildSystem` project that you created in previous recipe. Add a Java file `MainActivity.java` under the `cookbook.chapter8.portingwithbuildsystem` package. This Java file simply loads the shared library `PortingWithBuildSystem`, and calls the native method `naCreateABmp`.

2. Add the `mylog.h` and `PortingWithBuildSystem.c` files under it. `PortingWithBuildSystem.c` implements the native method `naCreateABmp`.

3. Create an `Android.mk` file under the `jni` folder to compile `PortingWithBuildSystem.c` as a shared library `libPortingWithBuildSystem.so`. The content of this `Android.mk` file is as follows:

```
LOCAL_PATH := $(call my-dir)
include $(CLEAR_VARS)
LOCAL_MODULE := libbmp-prebuilt
LOCAL_SRC_FILES := libbmp-0.1.3/lib/libbmp.a
LOCAL_EXPORT_C_INCLUDES := $(LOCAL_PATH)/libbmp-0.1.3/include/
include $(PREBUILT_STATIC_LIBRARY)
include $(CLEAR_VARS)
LOCAL_MODULE    := PortingWithBuildSystem
LOCAL_SRC_FILES := PortingWithBuildSystem.c
LOCAL_STATIC_LIBRARIES := libbmp-prebuilt
LOCAL_LDLIBS := -llog
include $(BUILD_SHARED_LIBRARY)
```

4. Add the `WRITE_EXTERNAL_STORAGE` permission to the `AndroidManifest.xml` file as follows:

```
<uses-permission android:name="android.permission.WRITE_EXTERNAL_
STORAGE"/>
```

5. Build and run the Android project. A bitmap file `test_bs_static.bmp` should be created at the `sdcard` folder of the Android device. We can use the following command to get the file:

```
$ adb pull /sdcard/test_bs_static.bmp .
```

The file is the same as `test_static.bmp` shown in the *Porting a library as shared library module with Android NDK build system* recipe of this chapter.

## How it works...

There are two common use cases for prebuilt libraries:

▶ You want to use a library from a third-party developer and only the library binary is provided

▶ You have already built a library and want to use the library without recompiling it

Your sample project belongs to the second case. Let's look at the things to consider when using a prebuilt library in Android NDK:

1. **Declare a prebuilt library module**: In Android NDK, a build module can either be a static or shared library. You have seen how a module is declared with source code. It is similar when a module is based on a prebuilt library.

    i. **Declare the module name**: This is done with the `LOCAL_MODULE` module description variable. In your sample project, define the module name with the following line:

    ```
    LOCAL_MODULE := libbmp-prebuilt
    ```

    ii. **List the source for the prebuilt library**: You will provide the path of the prebuilt library to the `LOCAL_SRC_FILES` variable. Note that the path is relative to `LOCAL_PATH`. In your sample project, list the path to the `libbmp.a` static library as follows:

    ```
    LOCAL_SRC_FILES := libbmp-0.1.3/lib/libbmp.a
    ```

    iii. **Export the library headers**: This is done through the `LOCAL_EXPORT_C_INCLUDES` module description variable. The variable ensures that any modules that depend on the prebuilt library module will have the path to the library headers appended to `LOCAL_C_INCLUDES` automatically. Note that this step is optional, as we can also explicitly add the path to library headers to any modules that depend on the prebuilt library module. However, it is a better practice to export the headers instead of adding the path on every module that depends on the prebuilt library module.

In your sample project, export the library headers with the following line in the `Android.mk` file:

```
LOCAL_EXPORT_C_INCLUDES := $(LOCAL_PATH)/libbmp-0.1.3/
include/
```

   iv. **Export the compiler and/or linker flags**: This can be done with `LOCAL_EXPORT_CFLAGS`, `LOCAL_EXPORT_CPPFLAGS`, and `LOCAL_EXPORT_LDLIBS`. This step is also optional and we won't use them in your sample project. You can refer to `docs/ANDROID-MK.html` at Android NDK for more detailed information about these module description variables.

   v. **Declare a build type**: You will need to include `PREBUILT_SHARED_LIBRARY` for the shared prebuilt library and `PREBUILT_STATIC_LIBRARY` for the static prebuilt library. In your sample project, use the following line to declare that you want to build a prebuilt static library module:

```
include $(PREBUILT_STATIC_LIBRARY)
```

2. **Use the prebuilt library module**: Once you have the prebuilt library module in place, you can simply list the module name in the `LOCAL_STATIC_LIBRARIES` or `LOCAL_SHARED_LIBRARIES` declaration of any module that depends on the prebuilt library. This is shown in your sample project's `Android.mk` file:

```
LOCAL_STATIC_LIBRARIES := libbmp-prebuilt
```

3. **Prebuilt library for debugging**: It is recommended by Android NDK that you provide the prebuilt library binaries that contain debug symbols, to facilitate debugging with `ndk-gdb`. When you package the library into an `apk` file, a stripped version created by Android NDK (at the project's `libs/<abi>/` folder) will be used.

We don't discuss how to generate the debug version of a library because it depends on how the library is built. Normally, the library documentation will contain instructions of how to generate a debug build. If you're building the library using GCC directly, then you can refer to `http://gcc.gnu.org/onlinedocs/gcc/Debugging-Options.html` for various options for debugging.

# Using a library in multiple projects with import-module

You may often need to use a library in multiple projects. You can put the library in each of the project's `jni` folders and build them separately. However, it is troublesome to maintain multiple copies of the same library. For example, when there is a new release of the library and you want to update the library, you will have to update each copy of the library.

Fortunately, Android NDK provides a feature to allow us maintain a library module outside a NDK project's main source tree and import the module with simple commands in the `Android.mk` file. Let's discuss how to import a module in this recipe.

## How to do it...

The following steps describe how to declare and import a module outside of a project's `jni` folder:

1. Create an Android application named `ImportModule` with native support. Set the package name as `cookbook.chapter8.importmodule`. Please refer to the *Loading native libraries and registering native methods* recipe of *Chapter 2, Java Native Interface* for more detailed instructions.

2. Add a Java file `MainActivity.java` under the `cookbook.chapter8.importmodule` package. This Java file simply loads the shared library `ImportModule`, and calls the native method `naCreateABmp`.

3. Download the `libbmp` library from `http://code.google.com/p/libbmp/downloads/list` and extract the archive file. Create a folder named `modules` under the project and a folder `libbmp-0.1.3` under the `modules` folder. Copy the `src/bmpfile.c` and `src/bmpfile.h` files from the extracted folder to the `libbmp-0.1.3` folder.

4. Follow step 3 of the *Porting a library as shared library module with Android NDK build system* recipe to update `src/bmpfile.h`.

5. Create an `Android.mk` file under the `libbmp-0.1.3` folder to compile `libbmp` as a static library `libbmp.a`. The content of this `Android.mk` file is as follows:

```
LOCAL_PATH := $(call my-dir)
include $(CLEAR_VARS)
LOCAL_MODULE := libbmp
LOCAL_SRC_FILES := bmpfile.c
LOCAL_EXPORT_C_INCLUDES := $(LOCAL_PATH)
include $(BUILD_STATIC_LIBRARY)
```

6. Add the `mylog.h` and `ImportModule.c` files to it. `ImportModule.c` implements the native method `naCreateABmp`.

7. Create an `Android.mk` file under the `jni` folder to compile `ImportModule.c` as a shared library `libImportModule.so`. The content of this `Android.mk` file is as follows:

```
LOCAL_PATH := $(call my-dir)
include $(CLEAR_VARS)
LOCAL_MODULE      := ImportModule
LOCAL_SRC_FILES := ImportModule.c
LOCAL_LDLIBS := -llog
LOCAL_STATIC_LIBRARIES := libbmp
include $(BUILD_SHARED_LIBRARY)
$(call import-add-path,$(LOCAL_PATH)/../modules)
$(call import-module,libbmp-0.1.3)
```

8. Add the `WRITE_EXTERNAL_STORAGE` permission to the `AndroidManifest.xml` file as follows:

```
<uses-permission android:name="android.permission.WRITE_EXTERNAL_
STORAGE"/>
```

9. Build and run the Android project. A bitmap file `test_bs_static.bmp` should be created at the `sdcard` folder of the Android device. You can use the following command to get the file:

```
$ adb pull /sdcard/test_im.bmp .
```

The file is the same as `test_static.bmp`, as shown in the *Porting a library as shared library module with Android NDK build system* recipe of this chapter.

## How it works...

In your sample project, you created a module outside the `jni` folder of the project and then imported the module for building the shared library `libImportModule.so`. The following steps should be performed when you declare and import a module:

1. **Declare the import module**: There is nothing special in declaring an import module. Since an import module is usually used by multiple NDK projects, it is a good practice to export the header files (with `LOCAL_EXPORT_C_INCLUDES`), compiler flags (`LOCAL_EXPORT_CFLAGS` or `LOCAL_EXPORT_CPPFLAGS`), and linker flags (`LOCAL_EXPORT_LDLIBS`) when declaring the import module.

   In our sample project, you declared an import static library module `libbmp`.

2. **Decide the place to put the import module**: The Android NDK build system will search the paths defined in `NDK_MODULE_PATH` for the import modules. By default, the `sources` folder of the Android NDK directory is appended to `NDK_MODULE_PATH`. Therefore, you can simply place our import module folder under the `sources` folder and the Android NDK build system will be able to find it.

Alternatively, you can place the import module folder anywhere and append the path to NDK_MODULE_PATH. In our sample project, place the import the libbmp module in the modules folder.

3. **Append the import path**: This is not needed when placing the import module folder under the sources directory of Android NDK. Otherwise, you will need to tell the Android NDK build system where the import modules are, by appending the path to NDK_MODULE_PATH. The import-add-path macro is provided by NDK to help you to append the path.

   In your sample project, you appended the modules folder to NDK_MODULE_PATH with the following line at jni/Android.mk:

   ```
   $(call import-add-path,$(LOCAL_PATH)/../modules)
   ```

4. **Import the module**: Android NDK provides a macro import-module to import a module. This macro accepts a relative path to the import module's folder where the Android.mk file of the import module is located. The Android NDK build system will search for all the paths defined at NDK_MODULE_PATH for the import module.

   In your sample project, you imported the module with the following line at the jni/Android.mk file:

   ```
   $(call import-module,libbmp-0.1.3)
   ```

   The NDK build system will search for the libbmp-0.1.3/Android.mk file at all NDK_MODULE_PATH directories for the import modules.

5. **Use the module**: Using an import module is just like using any other library module. You will need to link to the library by listing it at LOCAL_STATIC_LIBRARIES for the static library import module and LOCAL_SHARED_LIBRARIES for the shared library import module.

For more information about how to import modules, you can refer to docs/IMPORT-MODULE. html at Android NDK.

# Porting a library that requires RTTI, exception, and STL support

The Android platform provides a C++ runtime library at /system/lib/libstdc++.so. This default runtime library does not provide C++ exception and RTTI. The support for a standard C++ library is also limited. Fortunately, Android NDK provides alternatives to the default C++ runtime library, which makes porting of a large number of existing libraries that require exception, RTTI, and STL support, possible. This recipe discusses how to port a C++ library that requires RTTI, exception, and STL support. You will widely use the boost library as an example.

## How to do it...

The following steps describe how to build and use the `boost` library for Android NDK:

1. Install a customized Android toolchain with the following command:

   ```
   $ANDROID_NDK/build/tools/make-standalone-toolchain.sh
   --platform=android-9 --install-dir=/tmp/my-android-toolchain
   ```

   This should install the toolchain at the `/tmp/my-android-toolchain` folder.

2. Create an Android application named `PortingBoost` with native support. Set the package name as `cookbook.chapter8.portingboost`. Please refer to the *Loading native libraries and registering native methods* recipe of *Chapter 2, Java Native Interface* for more detailed instructions.

3. Add a Java file `MainActivity.java` under the `cookbook.chapter8.portingboost` package. This Java file simply loads the shared library `PortingBoost`, and calls the native method `naExtractSubject`.

4. Download the boost library from `http://sourceforge.net/projects/boost/files/boost/`. In this recipe, you will build the `boost` library 1.51.0. Extract the downloaded archive file to the `jni` folder. This will create a folder named `boost_1_51_0` under the `jni` folder as follows:

5. In a command line shell, go to the `boost_1_51_0` directory. Enter the following command:

```
$ ./bootstrap.sh
```

6. Edit the `user-config.jam` file under the `jni/boost_1_51_0/tools/build/v2` folder. Append the following content to the end of the file. You can refer to `http://www.boost.org/boost-build2/doc/html/bbv2/overview/configuration.html` for more information about boost configuration:

```
NDK_TOOLCHAIN = /tmp/my-android-toolchain ;
using gcc : android4.6 :
    $(NDK_TOOLCHAIN)/bin/arm-linux-androideabi-g++ :
    <archiver>$(NDK_TOOLCHAIN)/bin/arm-linux-androideabi-ar
    <ranlib>$(NDK_TOOLCHAIN)/bin/arm-linux-androideabi-ranlib
    <compileflags>--sysroot=$(NDK_TOOLCHAIN)/sysroot
    <compileflags>-I$(NDK_TOOLCHAIN)/arm-linux-androideabi/include/c++/4.6
    <compileflags>-I$(NDK_TOOLCHAIN)/arm-linux-androideabi/include/c++/4.6/arm-linux-androideabi
    <compileflags>-DNDEBUG
    <compileflags>-D__GLIBC__
    <compileflags>-DBOOST_FILESYSTEM_VERSION=3
    <compileflags>-lstdc++
    <compileflags>-mthumb
    <compileflags>-fno-strict-aliasing
    <compileflags>-O2
        ;
```

7. Try building the `boost` library with the following command:

```
$ ./b2 --without-python --without-mpi  toolset=gcc-android4.6
link=static runtime-link=static target-os=linux --stagedir=android
> log.txt &
```

This command will execute the `boost` build in the background. You can monitor the build output by using the following command:

```
$ tail -f log.txt
```

The build will take some time to finish. It will fail to build some targets. We can examine the errors in the `log.txt` file.

The first error is that the `sys/statvfs.h` file is not found. You can fix this by updating the `libs/filesystem/src/operations.cpp` file. The updated parts are highlighted as follows:

```
#    include <sys/types.h>
#    include <sys/stat.h>
```

```
#    if !defined(__APPLE__) && !defined(__OpenBSD__) && !defined(__
ANDROID__)
#       include <sys/statvfs.h>
#       define BOOST_STATVFS statvfs
#       define BOOST_STATVFS_F_FRSIZE vfs.f_frsize
#    else
#       ifdef __OpenBSD__
#          include <sys/param.h>
#       elif defined(__ANDROID__)
#            include <sys/vfs.h>
#       endif
#       include <sys/mount.h>
#       define BOOST_STATVFS statfs
#       define BOOST_STATVFS_F_FRSIZE    static_cast<boost::uintmax_
t>(vfs.f_bsize)
#    endif
```

The second error is that the `bzlib.h` file is not found. This is because `bzip` is available on Android. You can disable `bzip` by adding the following line at the top of `jni/boost_1_51_0/tools/build/v2/user-config.jam`:

```
modules.poke : NO_BZIP2 : 1 ;
```

The third error is that `PAGE_SIZE` is not declared in this scope. You can fix this by adding the following line to `boost_1_51_0/boost/thread/thread.hpp` and `boost_1_51_0/boost/thread/pthread/thread_data.hpp`:

```
#define PAGE_SIZE sysconf(_SC_PAGESIZE)
```

8. Try building the library again with the same command in step 5. This time the library will build successfully.

9. Add the `mylog.h` and `PortingBoost.cpp` files under the `jni` folder. The `PortingBoost.cpp` file contains the implementation for the native method `naExtractSubject`. The function will match each line of the input string `pInputStr` with a regular expression using the `boost` library's `regex_match` method:

```
void naExtractSubject(JNIEnv* pEnv, jclass clazz, jstring
pInputStr) {
    std::string line;
    boost::regex pat( "^Subject: (Re: |Aw: )*(.*)" );
    const char *str;
    str = pEnv->GetStringUTFChars(pInputStr, NULL);
    std::stringstream stream;
    stream << str;
    while (1) {
        std::getline(stream, line);
```

```
        LOGI(1, "%s", line.c_str());
        if (!stream.good()) {
          break;
        }
        boost::smatch matches;
        if (boost::regex_match(line, matches, pat)) {
            LOGI(1, "matched: %s", matches[0].str().c_str());
        } else {
          LOGI(1, "not matched");
        }
    }
}
```

10. Add an `Android.mk` file under the `jni` folder with the following content:

```
LOCAL_PATH := $(call my-dir)
include $(CLEAR_VARS)
LOCAL_MODULE := boost_regex
LOCAL_SRC_FILES := boost_1_51_0/android/lib/libboost_regex.a
LOCAL_EXPORT_C_INCLUDES := $(LOCAL_PATH)/boost_1_51_0
include $(PREBUILT_STATIC_LIBRARY)
include $(CLEAR_VARS)
LOCAL_MODULE     := PortingBoost
LOCAL_SRC_FILES := PortingBoost.cpp
LOCAL_LDLIBS := -llog
LOCAL_STATIC_LIBRARIES := boost_regex
include $(BUILD_SHARED_LIBRARY)
```

11. Add an `Application.mk` file under the `jni` folder with the following content:

```
APP_STL := gnustl_static
APP_CPPFLAGS := -fexceptions
```

12. Build and run the project. You can monitor the logcat output with the following command:

```
$ adb logcat -v time PortingBoost:I *:S
```

The following is a screenshot of the logcat output:

```
11-11 22:41:16.204 I/PortingBoost( 7191): Subject: this is a boost test
11-11 22:41:16.214 I/PortingBoost( 7191): matched: Subject: this is a boost test
11-11 22:41:16.214 I/PortingBoost( 7191): Hello, boost
11-11 22:41:16.214 I/PortingBoost( 7191): not matched
```

## How it works...

In your sample project, you first built the boost library using the Android toolchain as a standalone compiler. You then used the `regex` library from `boost` as a prebuilt module. Note that the `boost` library requires support for C++ exceptions and STL. Let's discuss how to enable support for these features at Android NDK.

**C++ runtime at Android NDK**: By default, Android comes with a minimal C++ runtime library at `/system/lib/libstdc++.so`. The library does not support most C++ standard library functions, C++ exceptions, and RTTI. Fortunately, Android NDK comes with additional C++ runtime libraries that we can use. The following table summarizes the features provided by different runtime libraries at NDK r8:

|  | **C++ standard library** | **C++ exceptions** | **C++ RTTI** |
|---|---|---|---|
| **system** | minimal | No | No |
| **gabi++** | minimal | No (yes if NDK r8d or later) | Yes |
| **stlport** | yes | No (yes if NDK r8d or later) | Yes |
| **gnustl** | yes | yes | Yes |

 The C++ exceptions have been added to `gabi++` and `stlport` since Android NDK r8d.

The system library refers to the default value that comes with the Android system. There is only a minimal C++ standard library support, and no C++ exceptions and RTTI. The C++ headers supported include the following:

```
cassert, cctype, cerrno, cfloat, climits, cmath, csetjmp, csignal,
cstddef, cstdint, cstdio, cstdlib, cstring, ctime, cwchar, new, stl_
pair.h, typeinfo, utility
```

▶ `gabi++` is a runtime library, which supports RTTI in addition to the C++ functions provided by the system default.

▶ `stlport` provides a complete set of C++ standard library headers and RTTI. However, C++ exception is not supported. In fact, Android NDK `stlport` is based on `gabi++`.

- ▶ gnustl is the GNU standard C++ library. It comes with a complete set of C++ headers, and supports C++ exceptions and RTTI.

The shared library file gnustl is named as libgnustl_shared.so instead of libstdc++.so in other platforms. This is because the name libstdc++.so is used by the system default C++ runtime.

The Android NDK build system allows us to specify the C++ library runtime to link to the Application.mk file. Based on the library type (shared or static) and which runtime to use, we can define APP_STL as follows:

| | Static library | Shared library |
|---|---|---|
| **gabi++** | gabi++_static | gabi++_shared |
| **stlport** | stlport_static | stlport_shared |
| **gnustl** | gnustl_static | gnustl_shared |

In your sample project, add the following line in Application.mk to use the gnustl static library:

```
APP_STL := gnustl_static
```

You can only link a static C++ library into a single shared library. If a project uses multiple shared libraries and all libraries link to a static C++ library, each shared library will include a copy of the library's code in its binary. This will cause some problems, because some global variables used by the C++ runtime library are duplicated.

The sources, headers, and binaries of these libraries can be found at the sources/cxx-stl folder of Android NDK. You can also refer to docs/CPLUSPLUS-SUPPORT.html for more information.

**Enable the C++ exception support**: By default, all C++ sources are compiled with `-fno-exceptions`. In order to enable C++ exceptions, you will need to choose a C++ library, which supports exceptions (`gnustl_static` or `gnustl_shared`), and do one of the following:

- At `Android.mk`, add exceptions to `LOCAL_CPP_FEATURES` as follows:

  `LOCAL_CPP_FEATURES += exceptions`

- At `Android.mk`, add `-fexceptions` to `LOCAL_CPPFLAGS` as follows:

  `LOCAL_CPPFLAGS += -fexceptions`

- At `Application.mk`, add the following line:

  `APP_CPPFLAGS += -fexceptions`

**Enable the C++ RTTI support**: By default, C++ sources are compiled with `-fno-rtti`. In order to enable the RTTI support, you will need to use a C++ library, which supports RTTI, and do one of the following:

- At `Android.mk`, add `rtti` to `LOCAL_CPP_FEATURES` as follows:

  `LOCAL_CPP_FEATURES += rtti`

- At `Android.mk`, add `-frtti` to `LOCAL_CPPFLAGS` as follows:

  `LOCAL_CPPFLAGS += -frtti`

- At `Application.mk`, add `-frtti` to `APP_CPPFLAGS` as follows:

  `APP_CPPFLAGS += -frtti`

# 9
# Porting an Existing Application to Android with NDK

In this chapter, we will cover the following recipes:

- ▶ Porting a command-line executable to Android with an NDK build system
- ▶ Porting a command-line executable to Android with an NDK standalone compiler
- ▶ Adding GUI to a ported Android app
- ▶ Using background threads at porting

## Introduction

The previous chapter covered various techniques of porting a native library to Android with NDK. This chapter discusses the porting of native applications.

We will first introduce how to build a native command-line application for Android with an Android NDK build system and the standalone compiler provided by NDK. We will then add a GUI for the ported application. Finally, we illustrate using a background thread to do the heavy processing and sending the progress update message from the native code to the Java UI thread for GUI updates.

We will use the open source Fugenschnitzer program throughout this chapter. It is a content-aware image resizing program based on the **Seam Carving** algorithm. The basic idea of this algorithm is to change the size of an image by searching for and manipulating the seams (a **seam** is a path of connected pixels from top to bottom, or left to right) from the original image. The algorithm is able to resize an image while trying to keep the important information. For readers who are interested in the program and the algorithm, refer to the project's main page at `http://fugenschnitzer.sourceforge.net/main_en.html` for more details. Otherwise, we can ignore the algorithm and focus on how the porting is done.

# Porting a command-line executable to Android with an NDK build system

This recipe discusses how to port a command-line executable to Android with an NDK build system. We will use the open source Fugenschnitzer program (`fusch`) as an example.

## Getting ready

You should read the *Porting a library as a static library with an Android NDK build* system recipe in *Chapter 8, Porting and Using Existing Libraries with Android NDK*, before going through this one.

## How to do it...

The following steps describe how to port the `fusch` program to Android with an NDK build system:

1. Create an Android application named **PortingExecutable** with native support. Set the package name as `cookbook.chapter9.portingexecutable`. Refer to the *Loading native libraries and registering native methods* recipe in *Chapter 2, Java Native Interface*, if you want more detailed instructions.

2. Remove the existing content under the `jni` folder of the project.

3. Download the source code of the `fusch` library and command-line application from `http://fugenschnitzer.sourceforge.net/main_en.html`. Extract the archive files and put them into the `jni/fusch` and `jni/fusch_lib` folders respectively.

4. Download `libpng 1.2.50` from `http://sourceforge.net/projects/libpng/files/libpng12/1.2.50/` and extract the files to the `jni/libpng-1.2.50` folder. The latest version of `libpng` won't work because the interface is different.

5. Add an `Android.mk` file under the `jni/libpng-1.2.50` folder to build `libpng` as a static library module. The file has the following content:

```
LOCAL_PATH := $(call my-dir)
include $(CLEAR_VARS)
LOCAL_CFLAGS :=
LOCAL_MODULE     := libpng
LOCAL_SRC_FILES :=\
  png.c \
  pngerror.c \
  pngget.c \
  pngmem.c \
  pngpread.c \
  pngread.c \
  pngrio.c \
  pngrtran.c \
  pngrutil.c \
  pngset.c \
  pngtrans.c \
  pngwio.c \
  pngwrite.c \
  pngwtran.c \
  pngwutil.c
LOCAL_LDLIBS := -lz
LOCAL_EXPORT_LDLIBS := -lz
LOCAL_EXPORT_C_INCLUDES := $(LOCAL_PATH)
include $(BUILD_STATIC_LIBRARY)
```

6. Add an `Android.mk` file under the `jni/fusch_lib` folder to build `libseamcarv` as a static library module. The file content is as follows:

```
LOCAL_PATH := $(call my-dir)
include $(CLEAR_VARS)
LOCAL_MODULE     := libseamcarv
LOCAL_SRC_FILES :=\
  sc_core.c  \
  sc_carve.c  \
  sc_color.c  \
  sc_shift.c \
  sc_mgmnt.c \
  seamcarv.c
LOCAL_CFLAGS := -std=c99
LOCAL_EXPORT_C_INCLUDES := $(LOCAL_PATH)
include $(BUILD_STATIC_LIBRARY)
```

7. Add the third `Android.mk` file under the `jni/fusch` folder to build the `fusch` executable, which uses the two static libraries built in the two folders `libpng-1.2.50` and `fusch_lib`.

```
LOCAL_PATH := $(call my-dir)
include $(CLEAR_VARS)
LOCAL_MODULE     := fusch
LOCAL_SRC_FILES := fusch.c
LOCAL_CFLAGS := -std=c99
LOCAL_STATIC_LIBRARIES := libpng libseamcarv
include $(BUILD_EXECUTABLE)
```

8. Add the fourth `Android.mk` file under the `jni` folder to include the `Android.mk` files under its subfolders.

```
LOCAL_PATH := $(call my-dir)
include $(call all-subdir-makefiles)
```

9. Build the application and you will see a binary file, `fusch`, under the `libs/armeabi` folder. We can put this binary into a rooted Android device or an emulator with the following command:

```
$ adb push fusch /data/data/
```

10. Note that we will not be able to copy and execute the binary on a non-rooted Android device because we cannot get the permission to execute.

11. Start the first command line on the console. We can grant the execution permission to the binary and execute it with the following command:

```
$ adb shell
```

```
# cd /data/data
```

```
# chmod 755 fusch
```

```
# ./fusch
```

This will print out the help message of the program.

12. Start the second command-line shell. Push the test PNG file `cookbook_ch9_test.png` (available under the `assets` folder of the sample project's source code) to the testing device or emulator with the following command:

```
$ adb push cookbook_ch9_test.png /data/data/
```

13. Get back to the first command-line shell and execute the `fusch` program again with the following command:

```
# ./fusch cookbook_ch9_test.png 1.png h-200
```

14. The program will take a while to resize the input image from 800 x 600 to 600 x 600. Once it is finished, we can get the processed image with the following command at the second command-line shell:

```
$ adb pull /data/data/1.png .
```

15. The following screenshot shows the original image on the left and the processed image on the right:

## How it works...

The sample project demonstrates how to port the `fusch` program as a command-line executable to Android. We describe the sources to the Android NDK build system in the `Android.mk` file and the NDK build system handles the rest.

The steps to port a command-line executable are as follows:

1. Figure out the library dependencies. In our sample program, `fusch` depends on `libseamcarv` (in the `fusch_lib` folder) and `libpng`, and `libpng` subsequently depends on `zlib`.

2. If a library is not available on the Android system, port it as a static library module. This is the case for `libseamcarv` and `libpng` in our sample application. But as `zlib` is available on Android, we simply need to link to it.

3. Port the executable as a separate module and link it to the library modules.

## Understanding the Android.mk files

We have covered most of the `Android.mk` variables and macros in *Chapter 8, Porting and Using Existing Libraries with Android NDK*. We will introduce two more predefined variables here. You can also refer to the Android NDK file `docs/ANDROID-MK.html` for information on more macros and variables.

- ▸ `LOCAL_CFLAGS`: A module description variable. This allows us to specify additional compiler options or macro definitions for building C and C++ source files. Another variable that serves a similar purpose is `LOCAL_CPPFLAGS`, but for C++ source files only. In our sample project, we passed `-std=c99` to the compiler when building `libseamcarv` and `fusch`. This asks the compiler to accept ISO C99 C language standard syntax. Failing to specify the flag will result in compilation errors at the time of building.

 It is also possible to specify the include paths with `LOCAL_CFLAGS += I<include path>`. However, it is recommended that we use `LOCAL_C_INCLUDES` because the `LOCAL_C_INCLUDES` path will also be used for `ndk-gdb` native debugging.

- ▸ `BUILD_EXECUTABLE`: A GNU make variable. It points to a build script that collects all information about the executable that we want to build and determines how to build it. It is similar to `BUILD_SHARED_LIBRARY` and `BUILD_STATIC_LIBRARY` except that it is for executables. It is used when building `fusch` in our sample project.

```
include $(BUILD_EXECUTABLE)
```

With this explanation and the knowledge we acquired in *Chapter 8, Porting and Using Existing Libraries with Android NDK*, it is now fairly easy to understand the four `Android.mk` files used in our sample application. We ported `libpng` and `libseamcarv` as two static library modules. We export the dependent libraries (with `LOCAL_EXPORT_LDLIBS`) and header files (with `LOCAL_EXPORT_C_INCLUDES`), so they are automatically included when using the module. When porting `libpng`, we also link to the `zlib` library (with `LOCAL_LDLIBS`) available on the Android system. Finally, we port the `fusch` program by referring to the two library modules (with `LOCAL_STATIC_LIBRARIES`).

# Porting a command-line executable to Android with an NDK standalone compiler

The previous recipe covered how to port a command-line executable to Android with an NDK build system. This recipe describes how to do it by using the Android NDK toolchain as a standalone compiler.

## Getting ready

It is recommended that you read the *Porting a library with its existing build system* recipe in *Chapter 8, Porting and Using Existing Libraries with Android NDK*, before continuing.

## How to do it...

The following steps describe how to port the `fusch` program to Android by using the NDK toolchain directly:

1. Create an Android application named **PortingExecutableBuildSystem** with native support. Set the package name as `cookbook.chapter9.portingexecutablebuildsystem`. Refer to the *Loading native libraries and registering native methods* recipe of *Chapter 2, Java Native Interface*, if you want more detailed instructions.

2. Remove the existing content under the `jni` folder of the project.

3. Download the source code of the `fusch` library and the command-line application from `http://fugenschnitzer.sourceforge.net/main_en.html`. Extract the archive files and put them into the `jni/fusch` and `jni/fusch_lib` folders respectively.

4. Download `libpng 1.2.50` from `http://sourceforge.net/projects/libpng/files/libpng12/1.2.50/` and extract the files to the `jni/libpng-1.2.50` folder. The latest version of `libpng` won't work because the interface has changed. Replace the `config.guess` script under `libpng-1.2.50` with the one at `http://gcc.gnu.org/svn/gcc/branches/cilkplus/config.guess` and `config.sub` with the script at `http://gcc.gnu.org/svn/gcc/branches/cilkplus/config.sub`.

5. Add a `build_android.sh` file under the `jni/libpng-1.2.50` folder to build `libpng`. The file has the following content:

```
#!/bin/bash
NDK=~/Desktop/android/android-ndk-r8b
SYSROOT=$NDK/platforms/android-8/arch-arm/
export CFLAGS="-fpic \
   -ffunction-sections \
   -funwind-tables \
   -D__ARM_ARCH_5__  -D__ARM_ARCH_5T__ \
   -D__ARM_ARCH_5E__  -D__ARM_ARCH_5TE__ \
  -Wno-psabi \
  -march=armv5te \
   -mtune=xscale \
   -msoft-float \
  -mthumb \
```

```
     -Os \
    -DANDROID \
     -fomit-frame-pointer \
     -fno-strict-aliasing \
     -finline-limit=64"
export LDFLAGS="-lz"
export CC="$NDK/toolchains/arm-linux-androideabi-4.4.3/prebuilt/
linux-x86/bin/arm-linux-androideabi-gcc --sysroot=$SYSROOT"
./configure \
    --host=arm-linux-androideabi \
    --prefix=$(pwd) \
    --exec-prefix=$(pwd) \
   --enable-shared=false \
   --enable-static=true
make clean
make
make install
```

6. Add a `build_android.sh` file under the `jni/fusch_lib` folder to build the library `libseamcarv`. The file content is as follows:

```
#!/bin/bash
NDK=~/Desktop/android/android-ndk-r8b
SYSROOT=$NDK/platforms/android-8/arch-arm/
export CFLAGS="-fpic \
   -ffunction-sections \
   -funwind-tables \
   -D__ARM_ARCH_5__  -D__ARM_ARCH_5T__ \
   -D__ARM_ARCH_5E__  -D__ARM_ARCH_5TE__ \
  -Wno-psabi \
  -march=armv5te \
   -mtune=xscale \
   -msoft-float \
  -mthumb \
   -Os \
   -fomit-frame-pointer \
   -fno-strict-aliasing \
   -finline-limit=64 \
  -std=c99 \
   -DANDROID "
export CC="$NDK/toolchains/arm-linux-androideabi-4.4.3/prebuilt/
linux-x86/bin/arm-linux-androideabi-gcc --sysroot=$SYSROOT"
AR="$NDK/toolchains/arm-linux-androideabi-4.4.3/prebuilt/
linux-x86/bin/arm-linux-androideabi-ar"
```

```
SRC_FILES="\
  sc_core.c  \
  sc_carve.c  \
  sc_color.c  \
  sc_shift.c \
  sc_mgmnt.c \
  seamcarv.c"
$CC $SRC_FILES $CFLAGS -c
$AR cr libseamcarv.a *.o
```

7. Add the third `build_android.sh` file under the `jni/fusch` folder to build the `fusch` executable, which uses the two static libraries built at the two folders `libpng-1.2.50` and `fusch_lib`.

```
#!/bin/bash
NDK=~/Desktop/android/android-ndk-r8b
SYSROOT=$NDK/platforms/android-8/arch-arm
CUR_D=$(pwd)
export CFLAGS="-fpic \
    -ffunction-sections \
    -funwind-tables \
    -D__ARM_ARCH_5__  -D__ARM_ARCH_5T__  \
    -D__ARM_ARCH_5E__  -D__ARM_ARCH_5TE__  \
   -Wno-psabi \
   -march=armv5te \
    -mtune=xscale \
    -msoft-float \
   -mthumb \
    -Os \
    -fomit-frame-pointer \
    -fno-strict-aliasing \
    -finline-limit=64 \
   -std=c99 \
   -DANDROID \
   -I$CUR_D/../fusch_lib \
   -I$CUR_D/../libpng-1.2.50/include"
export LDFLAGS="-Wl,--no-undefined -Wl,-z,noexecstack -Wl,-z,relro
-Wl,-z,now -lz -lc -lm -lpng -lseamcarv -L$CUR_D/../fusch_lib
-L$CUR_D/../libpng-1.2.50/lib"
export CC="$NDK/toolchains/arm-linux-androideabi-4.4.3/prebuilt/
linux-x86/bin/arm-linux-androideabi-gcc --sysroot=$SYSROOT"
SRC_FILES="fusch.c"
$CC $SRC_FILES $CFLAGS $LDFLAGS -o fusch
```

8. Build the two libraries `libpng` and `libseamcarv` and the `fusch` executable by executing the `build_android.sh` script in the three subfolders `libpng-1.2.50`, `fusch_lib`, and `fusch`. We shall find `libpng.a` under the `libpng-1.2.50/lib` folder, `libseamcarv.a` under the `fusch_lib` folder, and the `fusch` executable under the `fusch` folder.

9. We can put the binary file `fusch` to a rooted Android device or an emulator with the following command:

```
$ cd <path to project folder>/PortingExecutableBuildSystem/jni/
fusch
```

```
$ adb push fusch /data/data/
```

10. Note that we will not be able to copy and execute the binary on a non-rooted Android device because we cannot get the permission.

11. Start the first command-line shell. We can grant the execution permission to the binary and execute it with the following command:

```
$ adb shell
```

```
# cd /data/data
```

```
# chmod 755 fusch
```

```
# ./fusch
```

12. This will print out the help message of the program.

13. Start the second command-line shell. Push the test PNG file `cookbook_ch9_test.png` (available under the `assets` folder of the sample project's source code) to the testing device or emulator with the following command:

```
$ adb push cookbook_ch9_test.png /data/data/
```

14. Get back to the first command-line shell and execute the `fusch` program again with the following command:

```
# ./fusch cookbook_ch9_test.png 1.png v-200
```

15. The program will take a while to resize the input image from 800 x 600 to 800 x 400. Once it is finished, we can get the processed image with the following command at the second command-line shell:

```
$ adb pull /data/data/1.png  .
```

16. The following figure shows the original image on the left and the processed image on the right:

## How it works...

The sample project shows how to port a command-line executable to Android by using the NDK toolchain as a standalone compiler.

The steps to port the executable are similar to those in the previous recipe where we used the Android NDK build system. The key here is to pass proper options to the standalone compiler.

### Porting libpng

`libpng` comes with its own build scripts. We can get a list of options to configure the building process with the following command:

```
$ ./configure -help
```

The compiler command, compiler flags, and linker flags can be configured with the environment variables `CC`, `CFLAGS`, and `LDFLAGS` respectively. In the `build_android.sh` script under the `libpng-1.2.50` folder, we set these variables to use the NDK compiler to build for the ARM architecture. For a detailed explanation of how to port a library, we can refer to the *Porting a library with its existing build system using Android NDK toolchain* recipe in *Chapter 8*, *Porting a Library with its Existing Build System*.

We will now cover a few compilation options. Since the Android NDK toolchain is based on GCC, we can refer to `http://gcc.gnu.org/onlinedocs/gcc/Option-Summary.html` for a detailed explanation of each option.

- `-fpic`: It generates position-independent code suitable for building a shared library.
- `-ffunction-sections`: This option asks the linker to perform optimizations to improve the locality of reference in the code.
- `-funwind-tables`: It generates static data for unwinding the call stack.

▶ `-D__ARM_ARCH_5__`, `-D__ARM_ARCH_5T`, `-D__ARM_ARCH_5E__`, `-D__ARM_ARCH_5TE`, `-DANDROID` defines `__ARM_ARCH_5__`, `__ARM_ARCH_5T`, `__ARM_ARCH_5E__`, `__ARM_ARCH_5TE`, and `ANDROID` as macro, with definition equal to 1. For example, `-DANDROID` is equivalent to `-D ANDROID=1`.

▶ `-Wno-psabi`: It suppresses the warning message about `va_list` and so on.

▶ `-march=armv5te`: It specifies the target ARM architecture as `ARMv5te`.

▶ `-mtune=xscale`: It tunes the performance of the code as it will be running on the xscale processor. Note that xscale is a processor name.

▶ `-msoft-float`: It uses software floating point functions.

▶ `-mthumb`: It generates code using the Thumb instruction set.

▶ `-Os`: It provides optimization for size.

▶ `-fomit-frame-pointer`: It helps avoid saving frame pointers in registers if possible.

▶ `-fno-strict-aliasing`: No strict aliasing rules can be applied. This prevents the compiler from unwanted optimizations.

▶ `-finline-limit=64`: It sets the limit size of functions that can be inlined as `64` pseudo instructions.

▶ `-std=c99`: It accepts `c99` standard syntax.

When the build executes successfully, we can find the `libpng.a` static library under the `libpng-1.2.50/lib` folder and the header files under the `libpng-1.2.50/include` folder.

 The Android NDK build system essentially figures out the proper compilation options for us and invokes the cross compiler for us. Therefore, we can learn the options to pass to the compiler from the NDK build system output. For example, we can invoke the command `ndk-build -B V=1` or `ndk-build -B -n` in the previous recipe to see how the NDK build system handles the building of `libpng`, `libseamcarv`, and `fusch`, and apply similar options in this recipe.

## Porting libseamcarv

`libseamcarv` comes with a Makefile but no configure file. We can either modify the Makefile or write a build script from scratch. Since the library only contains a few files, we will write the build script directly. There are two steps to be followed:

1. Compile all source files to object files. This is done by passing the `"-c"` option at compilation.

2. Archive the object files into a static library. This step is done with the archiver `arm-linux-androideabi-ar` from the NDK toolchain.

As we have explained in *Chapter 8*, *Porting and Using Existing Libraries with Android NDK*, a static library is simply an archive of object files, which can be created by the `archiver` program.

### Porting fusch

We need to link to the two libraries we built, namely `libpng` and `libseamcarv`. This is done by passing the following options to the linker:

```
-lpng -lseamcarv -L$CUR_D/../fusch_lib -L$CUR_D/../libpng-1.2.50/lib
```

This "`-L`" option adds `fusch_lib` and `libpng-1.2.50/lib` to the library's search path and "`-l`" tells the linker to link to the `libpng` and `libseamcarv` libraries. The build script will output a binary file named `fusch` under the `fusch` folder.

The `fusch` program is fairly simple. Therefore, we can use either the Android NDK build system or a standalone compiler to port it. If an application has more dependencies, it can be difficult to describe everything in `Android.mk` files. Therefore, it is helpful that we can use the NDK toolchain as a standalone compiler and make use of a library's existing build scripts.

## Adding GUI to a ported Android app

The previous two recipes demonstrate how to port a command-line executable to Android. Needless to say, the biggest disadvantage of such a method is that it cannot be executed on a non-rooted Android device. This recipe discusses how to address the issue by adding a GUI when porting an application to Android.

### How to do it...

The following steps describe how to add a simple UI to the ported app:

1. Create an Android application named `PortingExecutableAUI` with native support. Set the package name as `cookbook.chapter9.portingexecutableaui`. Refer to the *Loading native libraries and registering native methods* recipe of *Chapter 2, Java Native Interface*, if you want more detailed instructions.

2. Follow steps 2 to 8 of the *Porting a command line executable to Android with NDK build system* recipe of this chapter.

3. Add a `mylog.h` file under the `jni/fusch` folder. Add the following lines to the `jni/fusch/fusch.c` file at the beginning of the main method, then remove the original main method signature line. The `naMain` method accepts a command from the Java code instead of the command-line shell. The arguments should be separated by a space:

```
#ifdef ANDROID_BUILD
#include <jni.h>
#include "mylog.h"
int naMain(JNIEnv* env, jclass clazz, jstring pCmdStr);

jint JNI_OnLoad(JavaVM* pVm, void* reserved) {
  JNIEnv* env;
  if ((*pVm)->GetEnv(pVm, (void **)&env, JNI_VERSION_1_6) != JNI_
OK) {
    return -1;
  }
  JNINativeMethod nm[1];
  nm[0].name = "naMain";
  nm[0].signature = "(Ljava/lang/String;)I";
  nm[0].fnPtr = (void*)naMain;
  jclass cls = (*env)->FindClass(env, "cookbook/chapter9/
portingexecutableaui/MainActivity");
  // Register methods with env->RegisterNatives.
  (*env)->RegisterNatives(env, cls, nm, 1);
  return JNI_VERSION_1_6;
}


 int naMain(JNIEnv* env, jclass clazz, jstring pCmdStr) {
   int argc = 0;
   char** argv = (char**) malloc (sizeof(char*)*4);
   *argv = "fusch";
   char** targv = argv + 1;
   argc++;
   jboolean isCopy;
    char *cmdstr = (*env)->GetStringUTFChars(env, pCmdStr,
&isCopy);
     if (NULL == cmdstr) {
       LOGI(2, "get string failed");
     }
     LOGI(2, "naMain assign parse string %s", cmdstr);
     char* pch;
```

```
    pch = strtok(cmdstr, " ");
    while (NULL != pch) {
      *targv = pch;
      argc++;
      targv++;
      pch = strtok(NULL, " ");
    }
    LOGI(1, "No. of arguments: %d", argc);
    LOGI(1, "%s %s %s %s", argv[0], argv[1], argv[2], argv[3]);
#else
  int main(int argc, char *argv[]) {
#endif
```

4. Add the following lines before the `return` statement of the main method to release the native string:

```
#ifdef ANDROID_BUILD
    (*env)->ReleaseStringUTFChars(env, pCmdStr, cmdstr);
#endif
```

5. Update the `Android.mk` file under `jni/fusch` as follows. The updated part is highlighted:

```
LOCAL_PATH := $(call my-dir)
include $(CLEAR_VARS)
LOCAL_MODULE    := fusch
LOCAL_SRC_FILES := fusch.c
LOCAL_CFLAGS := -std=c99 -DANDROID_BUILD
LOCAL_STATIC_LIBRARIES := libpng libseamcarv
LOCAL_LDLIBS := -llog
include $(BUILD_SHARED_LIBRARY)
```

6. Add the `MainActivity.java` file under the `cookbook.chapter9.portingexecutableui` package. The Java code sets up the GUI, loads the shared library `libfusch.so`, and calls the native method `naMain`.

7. Add an `activity_main.xml` file under the `res/layout` folder to describe the GUI.

8. In the `AndroidManifest.xml` file, add the following line before `<application>...</application>`:

```
<uses-permission android:name="android.permission.WRITE_EXTERNAL_
STORAGE"/>
```

9.  Build and run the Android app. You should be able to see a GUI similar to the following screenshot:

10. We can press either the **Width** or **Height** button to process the default image. Alternatively, we can load another `.png` image and process it. Once we click on either **Width** or **Height**, the GUI will become unresponsive and we will have to wait for the processing to finish. If the famous **Application Not Responding** (**ANR**) dialog box pops out, simply click on **Wait**.

11. When the processing finishes, the processed image will load and its dimensions will be displayed. The screenshot on the left shows the result for hitting the **Width** button, while the right one indicates the result for **Height** processing. Note that the images are scaled to fit into the display:

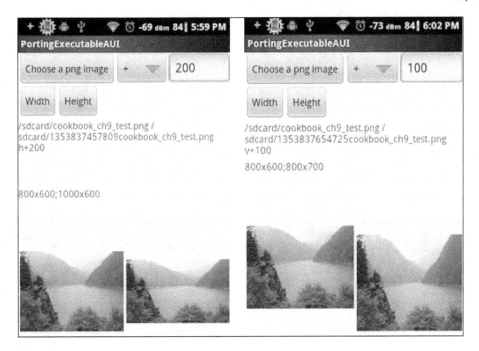

## How it works...

The example demonstrates how to add a GUI for the `fusch` program that we ported to Android. The `fusch` source code is modified for the native code to interface with the GUI.

In general, the steps can be followed to add a GUI to a command-line executable ported to Android.

1. Replace the main method with a native method. In our sample application, we replaced main with `naMain`.

2. Parse the input argument of the native method for command options instead of reading them from the command line. In our sample application, we parsed the third input argument `pCmdStr` for `fusch` command options. This allows the command to be constructed at the Java code and pass it easily to the native code .

3. Register the native method with the Java class.

4. In the Java code, the GUI can take various argument values specified by user, construct the command, and pass it to the native method for processing.

Note that in our modified native code, we didn't remove the original code. We used the C preprocessor macro `ANDROID_BUILD` to control which part of the source code should be included for building Android shared libraries. We pass `-DANDROID_BUILD` to the compiler in the `Android.mk` file (under the `fusch` folder), in order to enable the code specific for Android. This approach allows us to easily add support for Android, without breaking the code for other platforms.

There are two serious limitations for the sample application in this recipe. Firstly, the main UI thread handles the heavy image processing, which causes the application to become unresponsive. Secondly, there is no progress update when the image processing is going on. The GUI is updated only when the image processing is done. We will address these issues in the next recipe.

# Using background threads at porting

The previous recipe adds a GUI to the ported `fusch` program with two issues left behind—unresponsiveness of the GUI and no progress update when processing is going on. This recipe discusses how to use a background thread to handle the processing and report the progress to the main UI thread.

## Getting ready

The sample program in this recipe is based on the program we developed in previous recipes of this chapter. You should go through them first. In addition, readers are recommended to reading the following recipes in *Chapter 2, Java Native Interface*:

▶ *Calling static and instance methods from the native code*

▶ *Caching jfieldID, jmethodID, and reference data to improve performance*

## How to do it...

The following steps describe how to use a background thread for heavy processing and report progress update to the Java UI thread:

1. Copy the `PortingExecutableAUI` project that we developed in the previous recipe to a folder named `PortingExecutableAUIAsync`. Open the project in the folder at the Eclipse IDE.

2. Add the following code to `MainActivity.java`:

   `handler`: An instance of the `handler` class handles the messages sent from background threads. It will update the GUI with the content of the message.

   ```
   public static final int MSG_TYPE_PROG = 1;
   public static final int MSG_TYPE_SUCCESS = 2;
   public static final int MSG_TYPE_FAILURE = 3;
   ```

```
Handler handler = new Handler() {
  @Override
  public void handleMessage(Message msg) {
    switch(msg.what) {
      case MSG_TYPE_PROG:
        String updateMsg = (String)msg.obj;
        if (1 == msg.arg1) {
          String curText = text1.getText().toString();
          String newText = curText.substring(0, curText.
lastIndexOf("\n")) + "\n" + updateMsg;
          text1.setText(newText);
        } else if (2 == msg.arg1) {
          text1.append(updateMsg);
        } else {
          text1.append("\n" + updateMsg);
        }
        break;
      case MSG_TYPE_SUCCESS:
        Uri uri = Uri.fromFile(new File(outputImageDir +
outputImgFileName));
        img2.setImageURI(uri);
        text1.append("\nprocessing done!");
        text2.setText(getImageDimension(inputImagePath) + ";" +
        getImageDimension(outputImageDir + outputImgFileName));
        break;
      case MSG_TYPE_FAILURE:
        text1.append("\nerror processing the image");
        break;
    }
  }
};
```

ImageProcRunnable: A private class of MainActivity implements the Runnable interface, which accepts the command string, calls the native method naMain, and sends the result message to the handler at the Java UI thread. An instance of this class will be invoked from a background thread:

```
private class ImageProcRunnable implements Runnable {
  String procCmd;
  public ImageProcRunnable(String cmd) {
    procCmd = cmd;
  }
  @Override
  public void run() {
    int res = naMain(procCmd, MainActivity.this);
```

```
      if (0 == res) {
        //success, send message to handler
        Message msg = new Message();
        msg.what = MSG_TYPE_SUCCESS;
        handler.sendMessage(msg);
      } else {
        //failure, send message to handler
        Message msg = new Message();
        msg.what = MSG_TYPE_FAILURE;
        handler.sendMessage(msg);
      }
    }
  }
}
```

`updateProgress`: This is a method to be called from native code through JNI. It sends a message to the handler at the Java UI thread:

```
public void updateProgress(String pContent, int pInPlaceUpdate) {
  Message msg = new Message();
  msg.what = MSG_TYPE_PROG;
  msg.arg1 = pInPlaceUpdate;
  msg.obj = pContent;
  handler.sendMessage(msg);
}
```

3.  Update the `fusch.c` source code.

4.  We cache the `JavaVM` reference in the `naMain` method, and get a global reference for the `MainAcitvity` object reference `pMainActObj`. The `fusch` program uses more than one background thread. We will need these references to call Java methods from those background threads:

```
#ifdef ANDROID_BUILD
int naMain(JNIEnv* env, jobject pObj, jstring pCmdStr, jobject
pMainActObj);
jint JNI_OnLoad(JavaVM* pVm, void* reserved) {
  JNIEnv* env;
  if ((*pVm)->GetEnv(pVm, (void **)&env, JNI_VERSION_1_6) != JNI_
OK) {
    return -1;
  }
  cachedJvm = pVm;
  JNINativeMethod nm[1];
  nm[0].name = "naMain";
  nm[0].signature = "(Ljava/lang/String;Lcookbook/chapter9/
portingexecutableaui/MainActivity;)I";
  nm[0].fnPtr = (void*)naMain;
```

```
  jclass cls = (*env)->FindClass(env, "cookbook/chapter9/
portingexecutableaui/MainActivity");
  (*env)->RegisterNatives(env, cls, nm, 1);
  return JNI_VERSION_1_6;
}
int naMain(JNIEnv* env, jobject pObj, jstring pCmdStr, jobject
pMainActObj) {
  char progBuf[500];
  jmethodID updateProgMID, toStringMID;
  jstring progStr;
  jclass mainActivityClass = (*env)->GetObjectClass(env,
pMainActObj);
  cachedMainActObj = (*env)->NewGlobalRef(env, pMainActObj);
  updateProgMID = (*env)->GetMethodID(env, mainActivityClass,
"updateProgress", "(Ljava/lang/String;I)V");
  if (NULL == updateProgMID) {
    LOGE(1, "error finding method updateProgress");
    return EXIT_FAILURE;
  }
  int argc = 0;
  char** argv = (char**) malloc (sizeof(char*)*4);
  *argv = "fusch";
  char** targv = argv + 1;
  argc++;
  jboolean isCopy = JNI_TRUE;
  char *cmdstr = (*env)->GetStringUTFChars(env, pCmdStr,
&isCopy);
  if (NULL == cmdstr) {
    LOGI(2, "get string failed");
    return EXIT_FAILURE;
  }
  char* pch;
  pch = strtok(cmdstr, " ");
  while (NULL != pch) {
    *targv = pch;
    argc++;
    targv++;
    pch = strtok(NULL, " ");
  }
  LOGI(1, "No. of arguments: %d", argc);
  LOGI(1, "%s %s %s %s", argv[0], argv[1], argv[2], argv[3]);
#else
 int main(int argc, char *argv[]) {
#endif
```

5. Add the following lines before the `return` statement of the `main` method to release the native string and the cached JavaVM reference to avoid memory leaks:

```
#ifdef ANDROID_BUILD
    (*env)->ReleaseStringUTFChars(env, pCmdStr, cmdstr);
    (*env)->DeleteGlobalRef(env, cachedMainActObj);
    cachedMainActObj = NULL;
#endif
```

6. To update the GUI, we send out a message to the Java code. We need to update the code used to produce output messages at various parts of the source file. The following is an example of this:

```
#ifdef ANDROID_BUILD
  progStr = (*env)->NewStringUTF(env, MSG[I_NOTHINGTODO]);
  (*env)->CallVoidMethod(env, pMainActObj, updateProgMID, progStr,
0);
#else
  puts(MSG[I_NOTHINGTODO]);
#endif
```

7. The `seam_progress` and `carve_progress` functions are executed by native threads started at `naMain`. We used the cached `JavaVM` reference `cachedJvm` and `MainActivity` object reference `cachedMainActObj` to get `jmethodID` of the `updateProgress` method defined at `MainActivity.java`:

```
#ifdef ANDROID_BUILD
  char progBuf[500];
  JNIEnv *env;
  jmethodID updateProgMID;
  (*cachedJvm)->AttachCurrentThread(cachedJvm, &env, NULL);
  jstring progStr;
  jclass mainActivityClass = (*env)->GetObjectClass(env,
cachedMainActObj);
  updateProgMID = (*env)->GetMethodID(env, mainActivityClass,
"updateProgress", "(Ljava/lang/String;I)V");
  if (NULL == updateProgMID) {
    LOGE(1, "error finding method updateProgress at seam_
progress");
    (*cachedJvm)->DetachCurrentThread(cachedJvm);
    pthread_exit((void*)NULL);
  }
#endif
```

8. We can then call the `updateProgress` method from `seam_progress` and `carve_progress`. This is shown in the code section extracted from the `carve_progress` function, as follows:

```
#ifdef ANDROID_BUILD
  sprintf(progBuf, "%6d %6d %3d%%", max, pro, lrintf((float)(pro *
100) / max));
  progStr = (*env)->NewStringUTF(env, progBuf);
  (*env)->CallVoidMethod(env, cachedMainActObj, updateProgMID,
progStr, 1);
#else
  printf("%6d %3d%% ", pro, lrintf((float)(pro * 100) / max));
#endif
```

9. Build and run the Android app. You should be able to see a GUI similar to the following screenshot:

10. We can hit the **Width** or **Height** button to start the processing. The left and middle screenshots show the processing in progress, while the right screenshot shows the results:

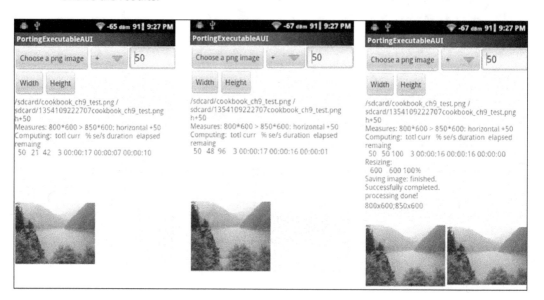

## How it works...

The preceding example shows how to use a background thread to handle heavy processing, so that the GUI can remain responsive to user inputs. While the background thread is processing the images, it also sends progress updates to the UI thread.

The details of the `fusch` program are actually a bit more complicated than the core idea described, because it uses heavy concurrent processing. This is illustrated in the following diagram:

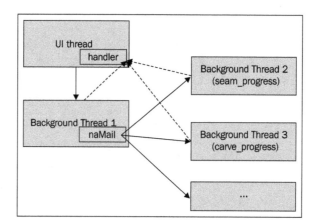

Once we click on either the **Width** or **Height** button in `MainActivity.java`, a new Java thread (**Background Thread 1**) will be created with an instance of the `ImageProcRunnable`. This thread will invoke the `naMain` native method.

Multiple native threads are created with the `pthread_create` function in the `naMain` method. Two of them, indicated as **Background Thread 2** and **Background Thread 3**, will be running `seam_progress` and `carve_progress` respectively.

We send messages of the `MSG_TYPE_PROG` type to the handler bound to the UI thread in all the three background threads. The handler will process the messages and update the GUI.

## Sending messages from the native code

Sending messages to a handler in Java is straightforward; we simply call the `handler.sendMessage()` method. But things can be a bit troublesome in the native code.

We defined an `updateProgress` method in `MainActivity.java`, which accepts a string and an integer, constructs a message, and sends it to the handler. The native code invokes this Java method through JNI in order to send messages. There are two situations:

- ▶ **Native code at Java thread**: This is the case for **Background Thread 1** in the previous diagram. The thread is created at Java code, and it calls the `naMain` native method. At `naMain`, we retrieve `jmethodID` for `updateProgress`, and call the `updateProgress` method through the JNI function `CallVoidMethod`. You can refer back to the *Calling static and instance methods from native code* recipe in *Chapter 2, Java Native Interface* for more information.

- ▶ **Native code at native thread**: This is what happens at **Background Thread 2** and **Background Thread 3**. These threads are created at `naMain` by the `pthread_create` function. We must call `AttachCurrentThread` to attach the native threads to a Java VM before we can make any JNI calls. Note that we used the cached `MainActivity` object reference `cachedMainActObj` for calling the `updateProgress` method. For more details about caching at JNI, we can refer to the *Caching jfieldID, jmethodID, and reference data to improve performance* recipe in *Chapter 2, Java Native Interface*.

The GUI we have created doesn't look all that good, but it is simple and enough to illustrate how to use a background thread for heavy processing and to send out GUI update messages from the native code.

# Index

## H

handle_activity_lifecycle_events function 182
handle_input_events 178
handler.sendMessage() method 317
HelloNDK app 89
Hello NDK program
  Java code 26
  native code 25
  native code, compiling 25
  writing, steps for 22-25

## I

IDE
  about 11
ImageProcRunnable class 311
import module
  appending 285
  declaring 284
  library, using in multiple projects 282-285
  placing 284
  using 285
import-module feature 266
initDisplay 155
initialElement 58
InlineAssemblyAddDemo method 81
Input Method Editor (IME) 180, 181
instance 165
instance field
  accessing 66
  accessing, in native code 62-67
instance methods
  calling 70
  calling, from native code 67, 68
instance object
  creating, in native code 54, 55
int 150
Integrated Development Environment. *See* IDE
interface 248
internalDataPath 164
Inter-process Communication (IPC) 172
int Java field type 64
int[] Java field type 64
int, Java type 38

IsAssignableFromDemo method 50
IsAssignableFrom JNI function 53
IsInstanceOf JNI function 56

## J

Java code 26, 28
javah
  using, steps for 26
Java JNI Specification
  URL 29
java.lang.System class 32
Java Native Interface. *See* JNI
Java primitive type mapping 38
Java program
  location 93
Java static
  accessing, in native code 62-67
Java string
  to native string 43
Java thread 51
jboolean variable 67
jfieldID data type 64
jfieldIDs 112
jintarray 60
JIT 7
jmethodID data type 68
jmethodIDs 72, 112
JNI
  about 28, 29
  arrays, manipulating 57-62
  assembly code, integrating 80-82
  character encoding 43
  classes, manipulating 50-53
  errors, checking 76-80
  exceptions, handling 76-80
  objects, manipulating 54-56
  Programmer' s Guide and Specification
    URL 29
  reference management 45, 46
  references, managing 43-45
  strings, manipulating 39-43
JNI character encoding 43
JNIEnv Interface Pointer 32, 33
JNIEnv pointer 25, 51
jnigraphics function 237

# P

parameters
  passing 34-37
PassingPrimitiveActivity.java Java code 36
passStringReturnString method 40
pDelete input parameter 47, 48
Perspective projection 132
policy
  scheduling 226
polling 203
porting
  background thread, using 310-315
POSIX Threads. *See* pthreads
prebuilt library
  build type, declaring 282
  compiler, exporting 282
  declaring 281
  for debugging 282
  headers, exporting 281
  library, using as 280-282
  linker flags, exporting 282
  module name, declaring 281
  source, listing 281
  using 282
  using, considerations for 281, 282
primitives
  about 124
  fragment, processing 124
  output, merging 124
  rasterization 124
  vertex processing 124
printf method 38
programmable pipeline 125
Projection transform 131, 132
pthread_cond_init function 203
pthread_cond_signal 204
pthread_cond_timedwait function 205
pthread_cond_timedwait_monotonic_np function 205
pthread_cond_timedwait_relative_np function 205
pthread_cond_timeout_np function 205
pthread_cond_wait 204
pthread_create function 194, 317
pthread_exit function 195
pthread_join function 195

pthread_key_create function 231
pthread_mutex_init function 200
PTHREAD_MUTEX_INITIALIZER macro 198
pthread_mutex_timedlock function 199
pthread_rwlock_destroy 209
pthreads
  about 158
  building, with 194
pthread_setspecific function 232

# R

rasterization 124
reader/writer lock
  attribute functions 212
  destroying 209
  initializing 209
  native threads synchronizing with, at Android NDK 206-209
  timed 211
  using 210, 211
readPng callback function 187
Realize() method 256
referenceAssignmentAndNew method 44
references
  about 113
  in JNI, manipulating 43-47
  managing, in JNI 45, 46
ReferenceTable overflow 47
RegisterNatives 33
ReleaseIntArrayElements JNI function 58, 61
release mode 113
renderAFrame, event type 154
renderThreadRun, event type 154
RenderView class 234
return statement 307, 314
Round Robin (RR) policy 227
RTLD_LAZY mode 240
RTLD_NOW mode 240
RTTI 96, 266
run_by_read_thread function 207, 210
run_by_thread1 function 201
run_by_thread2 function 196, 202
run_by_thread function 193, 194, 232
run_by_write_thread function 208
Run-time Type Information. *See* RTTI

# S

sampler 150
SCHED_FIFO 227
SCHED_OTHER 227
SCHED_RR 227
scheduling
  nice value/level used 227
sdkVersion 165
seam 294
Seam Carving 294
seam_progress function 314, 317
semaphores
  destroying 216
  initializing 216
  native threads synchronizing with, at Android
      NDK 212, 214
  operations 215
  using 216, 217
sensors
  accessing, at Android NDK 181-185
  accessing, steps for 184
  configuring 185
  default sensor of given type, getting 184
  disabling 185
  ensbling 185
  events, handling 185
  queue, creating 184
  reference, getting 184
SetIntArrayRegion function 60
setName method 71
SetObjectArrayElement 59
setpriority
  calling 228
SetPriority.cpp file 220
setValue method 71, 72
shader
  creating 151
  using, steps for 151
shared library module
  about 269
  library, porting as 266-269
  versus static 274, 275
short Java field type 64
short, Java type 38
SIMD 8, 98

Single Instruction Multiple Data. *See* SIMD
SLEngineItf interface 254
SLObjectItf interface 254, 256
specular light 138
start parameter 42
startPlaying 251
startRecording function 257
start_routine function 195
static
  calling, from native code 67, 68
  fields, accessing 65
static library module
  about 273
  library, porting as 271-274
  versus shared 274, 275
static methods
  calling 69
stdio function 245
stlport 290
String getName(), Java method 69
string Java field type 64
string length 43
strings
  in JNI, manipulating 39-43
surrogate pair 39
synchronizing
  at Android NDK 191
  native threads, with conditional variables
      200-203
  native threads, with mutex 1950197
  native threads, with reader/writer locks
      206-209
  native threads, with semaphore 212
sysroot directory 278
System.load method 32
System Properties window 9

# T

texture
  environment, setting 144
  filtering 144
  filtering, setting 144
  mapping 144
  mapping to 3D objects, with OpenGL ES 1.x
      API3D 140-144

wrapping 144
wrapping, setting 144
**texture coordinates 143**
**threads**
creating 194
priority, scheduling 226, 227
pthreads, building with 194
terminating 195
**thread-specific data key**
creating 231
**thread_step_2 function 230**
**ThrowNew function 79**
**toolchains 92**
**toString method 53**
**touch event handling 137**
**transformation, OpenGL ES 1.0**
ModelView transform 131
Projection transform 131
stages 131
Viewpoint transform 131
**trylock**
timed 211
**type safety 113**

# U

**Ubuntu Linux**
Android NDK development environment,
requisites for 16
Android NDK development environment,
setting up 16-18
Android NDK, updating 21
**Unicode Standard**
defining 39
**Unicode Transformation Format (UTF) 39**
**uniform 150**
**Universal Character Set (UCS) encoding 39**
**UNIX Development option 19**
**updateProgress method 312, 315, 317**
**UTF-8 string 76**
**UTF-16 string 76**

# V

**varying 150**
**Vector Floating Point (VFP) 98**
**vertex processing 124**
**vertex shader 149**

**VFPv3-D16 98**
**VFPv3-D32 98**
**Viewpoint transform 131**
**VM 7**
**vm attribute 164**
**void setName(String pName), Java method 69**

# W

**weak reference**
about 46, 48
versus global reference 46
versus local reference 46
**weakReference method 44**
**winant installer**
URL for installing 84
**window management**
about 158
ANativeWindow_fromSurface 158
ANativeWindow_setBuffersGeometrye 158
**Windows**
Android NDK development environment,
requisites for 9
Android NDK development environment,
setting up 9,-15
Android NDK, updating 20
Java JDK 6, requisite 9
**WRITE_EXTERNAL_STORAGE permission 268**

# X

**x86 ABI 97**
**x86 CPU family, CPU feature detections**
ANDROID_CPU_X86_FEATURE_MOVBE 104
ANDROID_CPU_X86_FEATURE_POPCNT 104
ANDROID_CPU_X86_FEATURE_SSSE3 104

# Z

**zipalign tool 91**
**zlib compression library**
programming with, in Android NDK 241-247
**ZlibDemo.cpp file 244**
**zlib library 247**

**Thank you for buying**
# Android Native Development Kit Cookbook

## About Packt Publishing

Packt, pronounced 'packed', published its first book "*Mastering phpMyAdmin for Effective MySQL Management*" in April 2004 and subsequently continued to specialize in publishing highly focused books on specific technologies and solutions.

Our books and publications share the experiences of your fellow IT professionals in adapting and customizing today's systems, applications, and frameworks. Our solution based books give you the knowledge and power to customize the software and technologies you're using to get the job done. Packt books are more specific and less general than the IT books you have seen in the past. Our unique business model allows us to bring you more focused information, giving you more of what you need to know, and less of what you don't.

Packt is a modern, yet unique publishing company, which focuses on producing quality, cutting-edge books for communities of developers, administrators, and newbies alike. For more information, please visit our website: www.packtpub.com.

## Writing for Packt

We welcome all inquiries from people who are interested in authoring. Book proposals should be sent to author@packtpub.com. If your book idea is still at an early stage and you would like to discuss it first before writing a formal book proposal, contact us; one of our commissioning editors will get in touch with you.

We're not just looking for published authors; if you have strong technical skills but no writing experience, our experienced editors can help you develop a writing career, or simply get some additional reward for your expertise.

# Android 4: New Features for Application Development

ISBN: 978-1-84951-952-6          Paperback: 166 pages

Develop Android applications using the new features of Android Ice Cream Sandwich

1. Learn new APIs in Android 4

2. Get familiar with the best practices in developing Android applications

3. Step-by-step approach with clearly explained sample codes

# AndEngine for Android Game Development Cookbook

ISBN: 978-1-84951-898-7          Paperback: 380 pages

Over 70 highly effective recipes with real-world examples to get to grips with the powerful capabilities of AndEngine and GLES 2

1. Step by step detailed instructions and information on a number of AndEngine functions, including illustrations and diagrams for added support and results

2. Learn all about the various aspects of AndEngine with prime and practical examples, useful for bringing your ideas to life

3. Improve the performance of past and future game projects with a collection of useful optimization tips

Please check **www.PacktPub.com** for information on our titles

# Java EE 5 Development with NetBeans 6

ISBN: 978-1-847195-46-3       Paperback: 400 pages

Develop professional enterprise Java EE applications quickly and easily with this popular IDE

1. Use features of the popular NetBeans IDE to improve Java EE development

2. Careful instructions and screenshots lead you through the options available

3. Covers the major Java EE APIs such as JSF, EJB 3 and JPA, and how to work with them in NetBeans

4. Covers the NetBeans Visual Web designer in detail

# ICEfaces 1.8: Next Generation Enterprise Web Development

ISBN: 978-1-847197-24-5       Paperback: 292 pages

Build Web 2.0 Applications using AJAX Push, JSF, Facelets, Spring and JPA

1. Develop a full-blown Web application using ICEfaces

2. Design and use self-developed components using Facelets technology

3. Integrate AJAX into a JEE stack for Web 2.0 developers using JSF, Facelets, Spring, JPA

Please check **www.PacktPub.com** for information on our titles